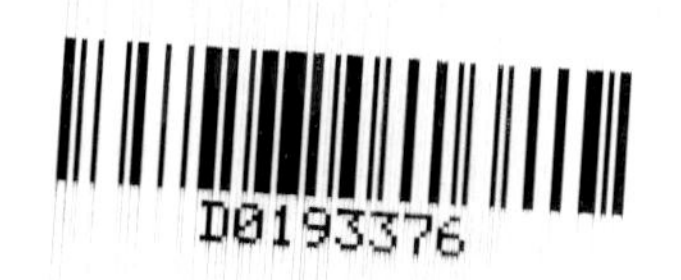

BOOJUMS ALL THE WAY THROUGH:

Communicating Science in a Prosaic Age

BOOJUMS ALL THE WAY THROUGH:

Communicating Science in a Prosaic Age

N. DAVID MERMIN
Professor of Physics, Cornell University

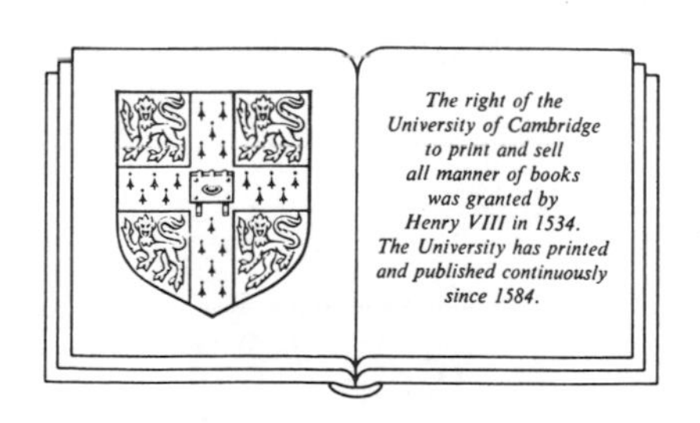
The right of the
University of Cambridge
to print and sell
all manner of books
was granted by
Henry VIII in 1534.
The University has printed
and published continuously
since 1584.

CAMBRIDGE UNIVERSITY PRESS
Cambridge
New York Port Chester Melbourne Sydney

Published by the Press Syndicate of the University of Cambridge
The Pitt Building, Trumpington Street, Cambridge CB2 1RP
40 West 20th Street, New York, NY 10011, USA
10 Stamford Road, Oakleigh, Melbourne 3166, Australia

First published 1990
Reprinted 1990

Printed in Great Britain at the University Press, Cambridge

British Library cataloguing in publication data

Mermin, N. David
Boojums all the way through: communicating science
in a prosaic age.
1. Scientific writing style
I. Title
808′0665

Library of Congress cataloguing in publication data

Mermin, N. David
Boojums all the way through: communicating science in a prosaic
age/N. David Mermin.
p. cm.
ISBN 0 521 38231 9. – ISBN 0 521 38880 5 (pbk.)
1. Physics. I. Title
QC71.M37 1990
530–dc20 89-37184 CIP

ISBN 0 521 38231 9 hardback
ISBN 0 521 38880 5 paperback

For Dorothy

CONTENTS

II. The quantum theory

III. Relativity

PREFACE

When I suggested to Simon Capelin that Cambridge University Press think about publishing a collection of my non-technical and pedagogical writings in and about physics, he responded politely, but expressed his concern that "a major problem is going to be trying to invent some sort of theme to the book."

"No need for invention," I thought. I have always felt that these essays and the aspect of my career as a physicist they embody have a greater unity and coherence than the rather diffuse and opportunistic conglomeration of my more technical efforts. But spelling this out has been a painful process. By putting my finger on the common thread of excitement and concern that led me to write them, I can't help feeling that I'm killing the magic some of these articles had (and still have) for me. I'd rather leave it to the reader to find (or fail to find) what I'm up to (or think I'm up to). But editors must be served, so here I go.

What the essays have in common is a concern with the problem of scientific communication. Some are addressed to a general audience, some to students, and some to other physicists; some are about physics, some about the tools of physics, and some about the practice of physics. But all share a preoccupation with both the substance and the style of written scientific communication.

Over the past fifty years or so, scientists have allowed the conventions of expression available to them to become entirely too confining. The insistence on bland impersonality and the widespread indifference to anything like the display of a unique human author in scientific exposition, have not only transformed the reading of most scientific papers into an act of tedious drudgery, but have also deprived scientists of some powerful tools for enhancing their clarity in communicating matters of great complexity. Scientists wrote beautifully through the 19th century and on into the early 20th. But somewhere after that, coincident with the explosive growth of research, the art of writing science suffered a grave setback, and the stultifying convention descended that the best scientific prose should sound like a non-human author addressing a mechanical reader.

All the essays in this collection are informed by my desire to cut through this atmosphere of verbal dreariness with which scientists seem unnecessarily to have surrounded themselves. My fervor is in no small part because physics, limply expressed, is less fun. Just as important, when we write physics or about physics as badly as we often now do, we undermine our science. We injure ourselves when we fail to make our discipline as clear and vibrant as we can to students – prospective scientists – and to the public who pay the taxes that support the science. I would even maintain that having acquired the habit of indifference to writing, we have made it easier for bad scientific writing to cover up bad thinking, thereby threatening science itself.

So although the subjects here range from reflections on the behavior of physicists to the philosophical implications of quantum mechanics, and although the essays vary in technical level from a commencement address to the construction and solution of a functional equation giving a new way to approach the relativistic mass–energy relation, they have these things in common: they take a new or offbeat look at a piece of everyday science or scientific practice with the aim of making it clearer or more vivid to scientists or non-scientists, and they write about it in a way that tries to avoid the pious, impersonal, heavy-handed style that has become one of the menaces of our professional life.

I

"*E pluribus boojum*" began as an after-dinner talk at a conference banquet. I had had too much wine, the projector fan kept blowing my transparencies away, Phil Anderson himself walked in at a critical moment, and the whole thing degenerated into utter pandemonium, to the point where a member of the band (scheduled next) was heard to mutter that it wasn't fair to ask them to go on after an act like that. I wrote it all up the next day for the *New Yorker* which, unlike any technical journal had the audacity to reject it without a word of explanation. Upgrading the level of the talk about physics, I sent it next to *Physics Today* – a wide-ranging publication received, whether they want it or not, by all members of the American Physical Society. *Physics Today* instantly understood what I was up to, provided some wonderful illustrations, and let it loose upon the community. A true tale of how physicists write their way toward immortality, "Boojum" caused a modest sensation, at least on the seismic scale of *Physics Today* articles. There was a torrent of letters to the editor, synopses appeared as far away as the *Sunday Times* in London, a footnote cited it in the next edition of the *Annotated Snark*, and a condensed version turned up a few years later in a collection of essays for a freshman composition course for budding scientists. I once encountered its aphoristic characterization of the refereeing process framed on a wall at Caltech.

I had another chance to try to convey my personal vision of the spirit of physics when I was asked to be the commencement speaker at St. John's College, the last of the great books schools. Surprised at the invitation, I asked, "did Carl Sagan say no and suggest me?" They were very nice about it. It was Lewis Thomas who had said no, then Czeslaw Milosz, and I was number three, apparently on the strength of "Quantum mysteries for anyone." I couldn't imagine what to say at such an occasion, but the temptation to be something as peculiar as a commencement speaker at a tiny liberal arts college in Santa Fe was irresistible. With great pain, I worked for several weeks on a speech on the subject of what a physicist could possibly have to say to the graduating class of the last of the great books schools, revising it right up to the morning of commencement. The graduates and their parents seemed to like it, so I sent the text off to *Physics*

Today on the theory that some of their readers might be curious what a physicist could possibly say to a bunch of great booksies. This time they disappointed me. "We've already had articles attacking Star Wars," they said.

The genesis of "My life with Landau", a personal memoir on the nature of scientific influence, is explained in the essay itself, the text of a talk I gave in Tel Aviv to honor L. D. Landau's memory on the 80th anniversary of his birth.[1] Technical details intrude here and there, but I would hope that the picture it reveals of how a great scientist can come to pervade an entire discipline, would still be accessible to the non-physicist reader, along with the *obiter dicta* on the nature of a Harvard education in the late 50s.

I had discovered at St. John's that it is possible actually to write out a talk in advance and then read it aloud. While a common practice among humanists, this procedure is virtually unheard of in physics, but I was coming to feel that it should not be dismissed out of hand, if you were as concerned with style as with substance. Sir Rudolph Peierls, a collaborator of Landau's and one of the great *raconteurs* of our profession, protested audibly from the audience when I announced my intention to proceed in this way. I said it was no worse than the prevailing rhetorical technique among physicists, reading the talk not from a manuscript but from sheets of plastic while projecting them on a screen, but Peierls fired back that he didn't like that either. So I compromised, reading my essay but sprinkling it with asides and encouraging the audience to interrupt. The high point of these exchanges occurred when I explained to the Israelis that "chutzpah" was a common American word that they might not be familiar with, thereby provoking cries of protest from the Soviet delegation that it was a common Russian word as well.

The five short "What's wrong with..." essays that conclude Part I appeared as columns in *Physics Today*. They all address various ways in which physicists fail to communicate with each other. The first four form a group, and come closer than anything

[1] The essay is preceded by a brief review of a biography of Landau, which conveys some of the flavor of the Landau legend.

else in this collection to expanding explicitly on the theme of communication in science, addressing various ways in which physicists fail to communicate with each other through the scientific journals. I am proud to announce here that *Physical Review Letters* is again spelling "Lagrangian" with an *i.* But whether anybody reads it any more is still an open question.

II

A major part of my efforts to explain physics to non-scientists, physics students, or other physicists, began when I agreed to talk about something – anything – to an informal gathering of Cornell faculty, almost none of them scientists. Having just learned of and being quite excited by Bell's theorem, I thought it might be fun to try to boil it down to something so simple that I could convey the argument using no mathematics beyond simple arithmetic and no quantum mechanics at all. To my pleasure, this turned out to be possible in principle, and perhaps even in practice too, since my audience seemed to get the point.

I described the argument to Ed Purcell, with whom I had been trading nuggets of physics pedagogy, and as an associate editor of the *American Journal of Physics*, the primary repository for articles on the teaching of physics, he urged me to write it up for them. That paper is not reproduced in this volume, though it did bring me the finest reward of my entire career in physics – a letter from Richard Feynman (coincidentally sent on my birthday) that begins "One of the most beautiful papers in physics that I know of is yours in the *American Journal of Physics*. . . ." But the *AJP* article turned out to be only the first step in a chain of papers. Purcell showed it to W. V. Quine, who urged me to produce a version for philosophers – the one reprinted here. The next variation further simplifies the *gedanken* experiment, by giving the argument a rather different twist. It was prepared for a conference of philosophers at Notre Dame that happened to take place on the last Saturday of the 1987 baseball season, as my beloved Mets were bowing out of the race. My most determinedly pedagogical effort appeared, of all places, in the 1988 annual update of the Encyclopaedia Britannica's Great Books. In addition to saying *what* happens, the essay in *Great Ideas Today* tries to convey to the non-technical reader *how* it is done, an exercise I had felt obliged to undertake ever since Viki Weisskopf reacted to the final sentence of my *AJP* paper ("It is left as a challenging exercise to the physicist reader to translate the elementary quantum-mechanical [explanation] into terms meaningful to a general reader...") by writing to say that it was my *duty* to provide such an explanation myself.

As is generally the case in physics, once you say anything with particular clarity you are treated as an expert on everything remotely related to the small point you have succeeded with great pain in mastering. People write, phone, or visit you seeking advice on matters of great intricacy and obscurity that you never pretended to understand. You are expected to deliver judgments on grant proposals in the area and referee for journals any submissions connected with what you have done, no matter how tenuous the link. You are invited to give talks (never mind that everything you know can be said in 15 minutes) at exotic watering spots, review books, and contribute to *Festschriften*. In the course of such efforts, no matter how valiantly you resist, you cannot avoid expanding your little island of competence, and by just such a process I was drawn from my simple device for explaining Bell's theorem into a somewhat broader consideration of quantum mysteries.

The essay on Bell's theorem I wrote for the commemoration of Bohr's 100th birthday differs from the version in *Great Ideas Today* by focusing on Bohr's reaction to the paper of Einstein, Podolsky, and Rosen that started the whole business. This was a somewhat tricky exercise, since I found and continue to find that reaction very hard to understand. My view of Bohr's position is expanded upon in a review of his philosophical writings, which strike me as frustratingly guarded. On the other hand I have little sympathy for Sir Karl Popper's great attack on Bohr, which though entertaining is fundamentally misguided. I conclude Part

II with another column from *Physics Today* addressing the different attitudes toward the quantum theory displayed by different generations of physicists. While the first four essays in this group attempt to communicate some very subtle aspects of quantum physics, the last three are more directly concerned with the problem of communication itself.

III

I began fighting the prevailing conventions of scientific exposition more or less by accident, in the midst of putting together an elementary introduction to the special theory of relativity, that I hoped might teach the subject at least as well as (in fact better than) the ordinary high school course propagates the more conventional body of physics. In a fit of revulsion at the bland and ponderous prose I had been slathering onto the page for several months, I suddenly found pouring out of me a mock Shakespearean drama that transformed my tedious cliches into ringing iambic pentameters that managed with a lunatic intensity to make the entirely technical point I had been struggling to express far more vividly than any of my earlier attempts, without in any way sacrificing its accuracy or compromising its subtlety. As a joke I sent the play to my publishers and to their everlasting credit McGraw Hill let me keep it in the book. Subsequently described as "the only Elizabethan drama that is explicitly Lorentz invariant", *Cruel Nature* was reprinted as (inexplicably) a guest *editorial* in *Physics Today*, performed at Christmastime in the Cavendish Laboratory with, I am told, Brian Josephson himself in the role of Friend of A, and put on the stage at Cornell (also at Christmas) with Ken Wilson putting in a cameo appearance as the Conductor of the chorus of relativists.

The remaining articles in the relativity group focus more on novel ways of teaching the subject than on novel literary forms for it. They assume somewhat more expertise on the part of the reader – an acquaintance with the elementary facts of special relativity – than do the essays on the quantum theory. "The amazing many colored Relativity Engine" (my private working title for the project, which the excellent editor of *AJP* [the successor of the man who fought against "Logarithms!" – see below] saw no reason to alter) describes a way to introduce students to the subject using no mathematics beyond simple arithmetic. Although the article is written as the description of a computer program, the pedagogical device that the program embodies was developed several years before the advent of the personal computer, and "sophisticated" readers might find the article of some interest in itself.

The Engine is followed by a brief article from *AJP* that derives simply and absolutely from scratch a result that is ordinarily encountered in expositions of relativity only as the fruit of applying a much more elaborately developed apparatus. Although this particular route to the velocity addition law *has* to have been noticed in the more than eighty years since 1905, I have yet to run across it in the relativity literature and, whether or not I ever do, I take pride in having come upon it independently all by myself.

Ed Purcell reacted to this argument by asking whether I couldn't get to the addition law with even fewer assumptions. It turned out he was right, but the route to the goal was quite a bit rockier. I include it ("Relativity without light") and the solution to the next Purcell challenge ("$E = mc^2$", worked out in collaboration with Mitch Feigenbaum) in spite of their heavier analytical load (a) because they still require nothing more than algebra and though most of this book is entirely accessible to those who do not speak algebra, I suspect that many people into whose hands it comes will have passed that barrier and (b) because both essays demonstrate the important fact that even venerably well established approaches to a subject can be reexamined and revised.

IV

"Logarithms!" addresses very directly the problem of communicating with students. Like all the relativity essays but the first, it embodies the solution to a technical problem: how to find a new and simpler formulation of a complex concept. It has a tone quite foreign to mathematical exposition, which I have never since been able to recreate to my satisfaction; the concluding Acknowledgment may offer some explanation for this. In addition to containing some arithmetical tricks that I had great fun constructing, "Logarithms!" also calls attention to the difficulty of trying to educate people who have had their curiosity crushed by dreadful mathematics courses and their interest in manipulating numbers destroyed by pocket calculators.

I had a devil of a time getting the essay published. The then editor of the *American Journal of Physics* refused even to send it to referees, insisting it had nothing to do with the teaching of physics. I said there wasn't much physics you could teach to people who couldn't cope with logarithms, but he was unmoved. I said my physicist friends thought it hit the nail right on the head; he said I didn't expect, did I, that my friends would come right out and say it was no good to my face. I actually visited him in his office at one point, a vast room filled with ancient wooden filing cabinets, and went through various futile arguments with him, not, mind you, trying to persuade him to publish it but merely to get the man to send it out for review. He wouldn't.

So I tried various journals of mathematics pedagogy. Their editors thought I was crazy – who would write an article about computing log 2 by hand in this day and age? Finally I found two physics texts which addressed the same problem in appendices, and, armed with these, persuaded at last the *AJP* editor to send the paper out to referees, who liked it. I was made to add a silly preface explaining why a reader of *AJP* might conceivably be interested, which I've left in the reprinting here because I like the way my irritation shines through. A few years after it appeared, "Logarithms!" was reprinted in the *American Mathematical Monthly*, I like to think for the moral improvement of mathematicians.

By the time I got around to writing the higher level sequel, "Stirling's formula!", *AJP* had a new editor who no longer considered me a loonie desperate for an audience, and he published it without a twitch. Although mathematically the most cumbersome piece in this collection, it manages to go surprisingly far, using little more than algebra (and a few calculus based embellishments at the end) toward elucidating the reasons behind a result that is ordinarily extracted with less intuitive insight and with very much more powerful mathematical machinery.

The tiny "Pi in the sky" was a letter to the editor of *AJP* responding to an earlier letter which had presented an extremely simple formula that reproduced to considerable accuracy a number traditionally calculated by the very elaborate methods of quantum electrodynamics, inviting readers to contemplate the likelihood that this could have happened by accident. I reproduce my note here in large part because of its title, which came to me very late and was substituted in stop-the-presses fashion.

I conclude with two brief reviews of treatises on mathematical subjects. My efforts to review technical books in a non-technical style have led to a few rather zany productions. Indeed, some time in the mid-seventies I noticed that *Physics Today* was no longer inviting me to review anything. The ban seems recently to have been lifted, but "What the hell are you doing and why the hell should I have anything to do with it?" has been the response, explicit or implicit, to many of my attempts to loosen up, sharpen, and enliven the process of communicating in and about physics. I hope that this volume, by bringing together many of these writings, will make their common intent clear.

But I won't stop even if it doesn't. I have, for example, a really neat solution to a homework problem in a Soviet physics book. An astronaut throws the lens cap of his camera away from his space capsule and you're asked some things about its subsequent orbit, which I found a nice way to answer using absolutely nothing but the conservation of the mysterious Runge–Lentz vector. I haven't written the article yet, but I have to do it because the title is too good to waste: "The lens cap and the Lentz vector."

I'm grateful to Dave Nicolaides for permission to reproduce his cartoon of the Boojum and the Dewar, emblematic of the 1980 Cornell Conference on Helium-3, and to Bob Richardson for rejuvenating with computer magic the only battered version of the drawing we could find. This collection owes more than she will believe to my mother, Eva G. Mermin, who always responds to my attempts to communicate science with "It's very impressive – I couldn't understand a single word," thereby inspiring me to ever greater efforts to be lucid; may she understand a word here and there. Anyone striving for a less somber approach to science could not ask for better colleagues than Geoffrey Chester, Michael Fisher, Kurt Gottfried, and Ben Widom; their wit and clarity have been a continuing inspiration, and their tolerance for my enthusiasms has been a great comfort. These virtues are shared by Neil Ashcroft, friend and letter writer, flower bringer and Goon Show fan, who kindly allowed me to sneak this collection in between the first and second editions of the world's funniest solid state physics text. John Rehr, Jason Ho, Paul Muzikar, Anupam Garg, David Wright, Sandra Troian, Dan Rokhsar, and Lisbeth Gronlund have all contributed, whether they knew it or not, to keeping me from drying up. But most important has been Dorothy Mermin, to whom this book is dedicated.

I.

Reflections on the pursuit of physics

1

E pluribus boojum: the physicist as neologist

I know the exact moment when I decided to make the word "boojum" an internationally accepted scientific term. I was just back from a symposium at the University of Sussex near Brighton, honoring the discovery of the superfluid[1] phases of liquid helium-3, by my Cornell colleagues Doug Osheroff, Bob Richardson, and Dave Lee. The Sussex Symposium took place during the drought of 1976. The Sussex downs looked like brown Southern California hills. For five of the hottest days England has endured, physicists from all over the world met in Sussex to talk about what happens at the very lowest temperatures ever attained.

Superfluid helium-3 is an anisotropic[2] liquid. The anisotropy is particularly pronounced in the phase known as ^{3}He-A. A family of lines weaves through the liquid ^{3}He-A which can be twisted, bent or splayed, but never obliterated by stirring or otherwise disturbing the liquid.

Several of us at the Sussex Symposium had been thinking about how the lines in ^{3}He-A would arrange themselves in a spherical drop of the liquid. The most symmetrical pattern might appear to have lines radiating outward from the center of the drop, like the quills of a (spherical) hedgehog (Fig. 1). There is an elegant topological argument, however, that such a pattern cannot be produced without at the same time producing a pair of vortex lines[3] connecting the point of convergence of the anisotropy lines to points on the surface of the drop (Fig. 2).

It appeared that if one did try to establish the symmetric pattern of radiating lines then the accompanying vortices would draw the point of convergence of the lines to the surface of the drop (Fig. 3), resulting in a final pattern that looked like Fig. 4.

When I returned to Ithaca I began to prepare for the proceedings the final text of the talk I had given which examined, among other things, the question of the spherical drop. Although no remarks about the drop were made after my talk, I decided to use the format of the discussion remark to describe the opinion that developed during the week that the symmetric pattern would collapse to one in which the lines radiated from a point on the surface. I found myself describing this as the pattern that remained after the symmetric one had "softly and suddenly vanished away." Having said that, I could hardly avoid proposing that the new pattern should be called a boojum.

The term "boojum" is from Lewis Carroll's "Hunting of the Snark" and it came to me at my typewriter rather as it had first come to Carroll as he walked in the country. The last line of a poem just popped into his head: "For the Snark was a Boojum, you see." A little distance along it was joined by the next to last line, "He had softly and suddenly vanished away." The hundreds of lines leading to this denouement followed in due course.

Goodness knows why "boojum" suggested softly and suddenly vanishing away to Carroll, but the connection having been made,

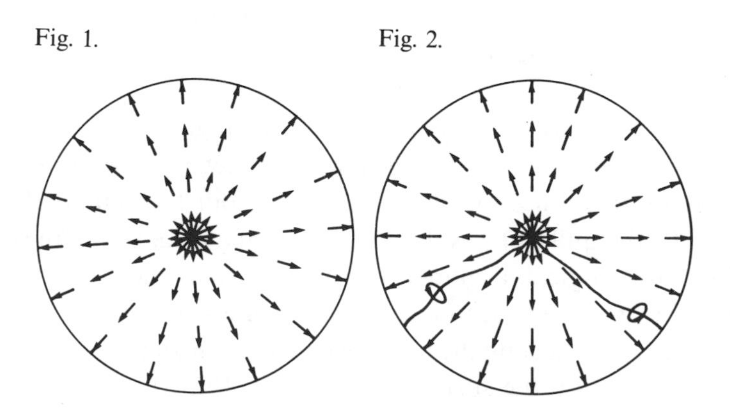
Fig. 1. Fig. 2.

it was inevitable that softly and suddenly vanishing away should suggest "boojum" to me. I was not unaware of how editors of scientific journals might view the attempt of boojums to enter their pages; I was not unmindful of the probable reactions of international commissions on nomenclature; nevertheless I resolved then and there to get the word into the literature.

There would be competition. Other people at the symposium had proposed calling my boojum a flower or a bouquet. Philip W. Anderson, who was to win his Nobel Prize the next year, and who appears at several critical moments in my tale, was not at the Sussex Symposium, but he was also thinking about spherical ^{3}He-A, and wanted to call the stable pattern a fountain. It is possible that Anderson's colleague at the Bell Telephone Laboratories, William F. Brinkman, was thinking about boojums at least a year before any of us. But he didn't know it was boojums he was thinking about and didn't bother to call them anything else either. Although he may not know it to this day, Brinkman was to play a role of pivotal importance in boojum's progress.

The first step toward assuring the adoption of "boojum" was easily accomplished. The discussion remarks, including my nomenclatural proposal in the guise of a discussion remark, received the usual minimal editorial attention. Shortly thereafter the word "boojum" – albeit encumbered by quotation marks – made its maiden appearance in the literature of modern physics.

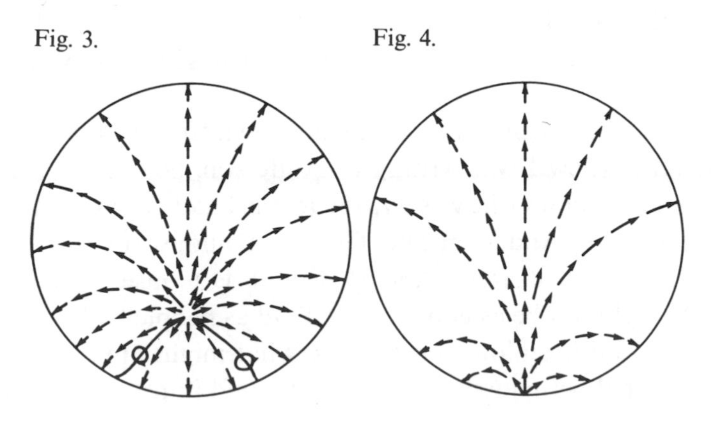

Fig. 3. Fig. 4.

> Maki and Hall (see also their prepared talks) both felt that the yozh would be unstable. They disagreed on the fashion in which it would disappear, but it was my impression that both favored the texture of fig. 3c as that prevailing after the yozh had softly and suddenly vanished away. Hall proposed calling it the "flower texture," but I personally think that "boojum" is more to the point.

Note my use of the word "yozh" to describe the symmetric pattern. "Yozh" is Russian for "hedgehog." I have never forgotten the word since studying Russian for the PhD language requirement, because it is only two letters long in Cyrillic. Since a Russian physicist, writing in Russian, had introduced the term, I used "yozh" instead of the increasingly popular "hedgehog" in the text of my paper. This effort to make an English word out of "yozh" failed utterly. Little did I then suspect that I would one day succeed in making a Russian word out of "boojum."

Published boojums

My next step was clearly to publish something which put my nomenclatural proposal to use, calling a boojum a boojum without fanfare or quotation marks. I wasn't ready to fight with editors of journals, but I was to deliver a paper on superfluid helium-3 at a conference to be held on Sanibel Island in January 1977, and the conference proceedings were to be published as a book. The form of my contribution to the Sanibel proceedings as well as the intensity of my interest in the boojum was considerably influenced by a series of letters I exchanged with Phil Anderson that fall.

Our correspondence was somewhat constrained by the fact that although I knew I was writing to Anderson, he – at least for a while – did not know he was writing to me. I had been sent the text of a paper by Anderson and Gerard Toulouse to referee for *Physical Review Letters*. Anderson and Toulouse argued that ^{3}He-A might not be as good a superfluid as people had expected, producing an ingenious reason why what might appear to be a conventional permanent supercurrent could in fact lose its flow.

I saw a possible flaw in their argument. The surface of a container has a rather peculiar effect on the anisotropy lines of ^{3}He-A: they are forced to line up perpendicular to the surface at the boundaries of the liquid (as they are in the pictures of the spherical drops shown above). Although this seemed to be of no relevance to the argument of Anderson and Toulouse, I worried that it might, in fact, invalidate the mechanism they proposed for the disappearance of the supercurrent. I suggested that such questions should be cleared up before the paper was published. In the only mildly acrimonious correspondence that ensued, the authors and I both started to realize that the ability of the surface to stabilize the supercurrent was indeed relevant, but that this stabilizing power could be lost if there was even a single boojum (none of us called it that) on the surface.

This was interesting enough for their paper to appear (though without the word boojum), and I found myself more committed than ever to establishing the term. It was now evident that boojums were more than an inert feature of the structure of ^{3}He-A drops; they had a vital role to play in the most fundamental property of the liquid, its superfluid flow. Furthermore it was no longer a bad idea about the pattern in spherical drops that was softly and suddenly vanishing away, but the supercurrent itself, whose soft and sudden vanishing could be triggered by a well placed boojum. My nomenclatural impulse had acquired the character of a prophetic vision.

In "The Hunting of the Snark," a boojum is a singular variety of snark with the alarming ability to bring about the soft and sudden vanishing away of anyone encountering it. The boojum in ^{3}He-A, being a point at which lines in different directions all meet, is a mathematical singularity. A singularity in ^{3}He-A responsible for the vanishing of a supercurrent *had* to be called a boojum.

Accordingly, at the appropriate point in my paper for the Sanibel Symposium, I let loose a flock of boojums:

> This twisted boojum is shown at the bottom of the torus in the cross section in which it resembles the hyperbolic

twistless boojum of Fig. 10 (but note that the cross section perpendicular to the page will resemble the circular boojum). If either of the booja encircles its part of the torus (see inset) then two quanta are subtracted from the circulation about the entire torus.

Inspection of this specimen reveals that I had adopted "booja" as the plural form. As we shall see this turned out to be a serious error. I believe it was the one false step I made in an otherwise impeccable campaign, though publishing this article may turn out to have been the second.

The editor of the proceedings approved "boojum"; indeed, he allowed it to appear in the index, where it can be seen in the company of better established but, I would have said, far less poetic technical terms in **B**:

The Sanibel Symposium took place during the remarkably cold winter of 1976–77. Many of the people who suffered through the heat of the Sussex Symposium found themselves together again, six months later, experiencing the coldest Florida January on record. They were rewarded by the first public lecture (mine) in which the word "boojum" was used in its new scientific context. (Anderson also spoke at Sanibel but he called the boojum a fountain.)

A boojum in Erice

Returning from cool Florida to frigid Ithaca, I set to work preparing a set of lecture notes that I delivered that June at a summer school in Erice, a mountain-top Sicilian town three thousand feet straight above the sea, whose streets and alleys are paved in geometrical patterns of massive stones, polished smooth by feet and wheels. The views in all directions were spectacular and the weather was neither too hot nor too cold. Occasionally a cloud would settle over the mountain top for a day or two sending cool mists swirling through the steep alleyways. The perfect place to meet a boojum, though nobody ever did.

The boojum, however, did make a casual but prolonged appearance in one of my lectures. The index entry in the published volume was worthy of a fully mature technical term:

> BCS gap, 175, 242, 244, 258
> Bogoliubov–Zubarev method, 274
> Boltzmann equation, 129
> Boojum, 214
>   circular, 223
>   hyperbolic, 223
>   and sound attenuation, 224
>   and superflow, 224
> Bose–Einstein condensation, 198
> Bose liquid, 36

A month later I talked about boojums at a conference in New Hampshire. No proceedings were published, but several Russians attended the meeting and it seemed important to get them thinking boojum too. I had hoped that the first person plural future form of the Russian verb to be ("budyem", more or less pronounced "BOODyum") would make them receptive to the term, but I was never then or thereafter able to convince any Russian that the two words resembled each other in any way. No matter. The Russians took to boojums at once, and one even said a boojum or two in his own talk. The weather was once more decidedly boojumish. I believe Concord recorded several of the hottest July days in its history. The temperature was 102 or 104, and Lake Winnipesaukee, in which many of us swam before, after, and sometimes during lectures, was as warm as a bath.

I returned from New Hampshire convinced that the boojum was ready to make its debut in an established scientific journal, but before I could consider how to bring this about there were two alarming developments.

Chia-Ren Hu, whom I had met at Sanibel, became interested in boojums, and wrote a paper, "Exact solution of a 'boojum' texture in ^{3}He-A." He sent it to the *Journal of Low Temperature Physics* and in due course received a letter from the editor, which read in part

> I have just received the comments of our referee on your paper and I enclose a copy of them. As you will see, he considers that the paper should be published provided the word "Boojum" be replaced with a suitable scientific word or phrase in the title, abstract, and text (p. 4, lines 8 & 17; p. 6, line 6). I too as General Editor concur unreservedly in this requirement. If you are willing to make such changes, we shall be happy to publish the paper.

Dr. Hu sent this communication on to me, together with a copy of the report of the referee who actually recommended only the removal of the boojum from the title. Dr. Hu informed me that it was his plan to substitute for the word "boojum" the acronym "SOSO" (for "singular on surface only"). Appalled at the imminent possibility of my boojum turning into a SOSO, I wrote immediately to the General Editor, dissenting unreservedly from his conclusions:

> Professor John G. Daunt
> Editor, Journal of Low Temperature Physics
>
> Dear Professor Daunt:
>
> I have received from C.-R. Hu a copy of a referee's report on his article on surface point singularities in superfluid ^{3}He-A, recommending that the word "boojum" be removed from the title, accompanied by a copy of a letter from you stipulating that the word be removed from the abstract and text as well. I agree that the term has not gained the currency to warrant its use in a title, but there is no reason for it to be excised from the text provided proper citations are given.
>
> The term "boojum" is now in wide use in the field of superfluid helium-3. It first appeared at the Sussex Symposium (August 1976) and can be found in print in *Physica*, **90B** + **C**, 1 (1977). The term appears again in the proceedings of the Sanibel Symposium (January 1977) to be published by Plenum within a month or two. There its literary origins are

explained and its use justified. Briefly, a boojum is a singular beast, the appearance of which causes the observer to "softly and suddenly vanish away." This is precisely the role played by the boojum in ^{3}He-A. If a boojum is in the container it can catalyze the decay of the supercurrent. Such a process is unique to ^{3}He-A and it requires a new nomenclature. The word "boojum" is sanctified by Webster's unabridged (2nd edn) where it is defined essentially as I have done above. It is therefore as respectable a term as the currently fashionable "hedgehog" (and rather more respectable than "quark").

In addition to the two citations mentioned above, the term "boojum" will appear in print in the proceedings of the June 1977 Erice Summer School on Quantum Liquids, to be published by North Holland this fall; it appears in a recent preprint of a *review* article by Brinkman and Cross; it appears in a recent preprint from the Landau Institute; and it was widely used and understood in discussions at last month's Gordon Conference on non-equilibrium phenomena in quantum liquids, attended by most of the world's experts on superflow in ^{3}He-A.

In short, "boojum" has now been used in the field for a year as a technical term meaning, quite precisely, "any surface point singularity the motion of which can catalyze the decay of a supercurrent." The term is specific, apt, and recognized by Webster. It has the virtue of being easily pronounced in Russian (since it is a homonym for the first person plural of the Russian verb to be). It is already in print in one reputable journal and will appear in print in at least five other places within half a year.

To ask that Dr. Hu resort to circumlocutions in the text of his article serves no linguistic or esthetic purpose, and obscures the physical significance of the point he wishes to make. Now that I have told you a little more about the meaning and widespread use of the term, may I urge you to let him reintroduce it in his text.

Sincerely yours,
N. David Mermin
Professor of Physics

The reference to the dictionary is important. If Hu's paper had mentioned a "flower texture" or a "fountain texture" chances are it would not have sounded alarms in the editorial office. It had occurred to me that if "boojum" were in the dictionary the character of the dispute would change. Not surprisingly, dictionaries readily at hand did not contain "boojum," but I did find it listed as an ordinary common noun in a copy I have at home of Webster's Unabridged, which I had won as an American history prize in high school in 1951. This was a few years before the appearance of the notorious 3rd edition, a fact of central importance in what was to follow. Had I not won the American history prize thirty years ago, "boojum" would not today be an internationally accepted scientific term.

My argument that "boojum" was a harmless common noun did not persuade Professor Daunt:

> My editorial Board and myself maintain a policy of asking our contributors to avoid the use of words or strings of letters in titles or in abstracts, the meanings of which may not be immediately recognised by our international reading subscribers and fellow physicists. Moreover we extend the policy to require contributors to define such words or letter complexes clearly if they wish to introduce them in the text of articles. There have been many occasions in the past when I have asked authors to accept this policy of ours and I assure you that Dr. Hu is by no means the first author to be requested to make changes in this regard.
>
> I myself was well aware of the meaning that you have attached to the word "Boojum" since, amongst other occasions, I was in your audience at Sanibel last year. I am, of course, aware of its origin. However, at the moment it is not only my opinion, but also that of the reviewers of Dr. Hu's paper, that the physico-technical meaning of the word is insufficiently known to the international audience of our Journal to warrant its use as Dr. Hu wished to use it in his paper.
>
> I look forward to the time when your new word may gain international acceptance and in the meantime I am maintaining active discussions about it with my Board of Editors and other reviewers of our Journal.

I was taken aback at the news that Daunt was actually present at Sanibel, carrying as it did the clear implication that he had seen through my bluff and bluster and knew as well as I did that all of my impressive array of published appearances of "boojum" were due to my authorship alone. Although I approved in general of the policy he was defending, I did feel that he and his editorial board were disappointingly unable to spot a good exception to the rule when one landed in their laps. I was also taken in by his argument about international readers. Only later did I learn that the boojum appears not only in "The Hunting of the Snark," but also in "La Chasse au Snark," "Die Jagd nach dem Schnark," "La caccia allo Snarco," "Snarkjakten," and "Snarkejagten," to name only a few. All I got out of the exchange was a certain quiet pleasure in trying to imagine some of the active discussions among the board of editors and reviewers.

Clearly the gauntlet had been thrown down. To his credit, Daunt was as ruthlessly impermissive with the odious SOSO as he had been with my gentle boojum, but it was now essential that "boojum" appear in print in the most authoritative and widely circulated of all the international journals of physics: *Physical Review Letters.*

Lexicographic complications

Of central importance to the success of this enterprise was the second alarming development. I received the text of a review of recent developments in the theory of superfluid helium-3 by W. F. Brinkman and M. C. Cross, both of Bell Labs. Leafing through the section of greatest interest to me, I was mortified to read

> the expected configuration for a sphere is the "boogum" shown in Fig. 3.

The careful reader of my letter to Daunt will notice that I tried to make the best of even a bad business like this, but here was clearly a new kind of trouble. The manuscript was being circulated prior to publication, but would assume an authoritative position when it did appear. It was essential to set the spelling straight. I thought I would simply cite Webster to Brinkman and, being in my office rather than at home, went to the nearby Physical Sciences Library to look up the exact citation. I was appalled to find:

> **boogum** or **boojum** (perh. fr. **boojum**, an imaginary creature in *The Hunting of the Snark* by Lewis Carroll (C. S. Dodgson) 1898 Eng. mathematician & writer; fr its grotesque appearance): a spiny tree (*Idria columnaris*) of the family Fouquieriacea chiefly of Lower California, sometimes arching over and rooting at its tips.

The library edition of Webster's was the third. Back at home I read again the lucid and concise second edition:

> **boojum**, n. In Lewis Carroll's *Hunting of the Snark*, a species of snark the hunters of which "softly and suddenly vanish away."

I compared the two editions on snark:

2nd edition:

> **snark**, n. (A blend of *snake* and *shark*). A nonsense creature invented by Lewis Carroll (Charles L. Dodgson), in his poem, *The Hunting of the Snark* (1876). One variety is known as the *boojum* (which see).

3rd edition:

> **snark** (prob alter of snork) dial Brit: snore, snort.

In some ways this was absolutely uncanny. One has to understand, to begin with, that physicists from Bell Labs have a celebrated and annoying habit of disagreeing with you and being right, or agreeing with you but getting there first. More than once had Brinkman casually corrected public slips of mine, or pointed out politely that some of my more beautiful thoughts had already been thought by him – as I would have realized had I troubled to read the proceedings of last year's (or even the year before's) Scottish Summer School, for example. If you look again at Fig. 4 and think of a tree, "arching over and rooting at its tips," you do indeed begin to wonder whether the boojum in ^{3}He-A shouldn't, in fact, be called a boogum instead. This was not to be the last time that the telephone company threatened to snatch my boojum away from me.

On the other hand, who could favor the entry in the 3rd edition? I took a firm line, writing Brinkman a letter of which I can find no copy. I do have his reply, which suggests that I must have struck him as raving incoherently about California trees and abominable revisions of once noble dictionaries. After some empty but soothing remarks he replied that "boogum" was a typist's slip which would be corrected before the article reached print. (It was not.)

As a result of this unfortunate episode I not only felt it essential to get "boojum" into the most distinguished of journals with the greatest possible speed, but also realized it would be necessary to ward off the upstart "boogum" while doing so.

There are two problems facing one wanting to get "boojum" into *Physical Review Letters.* The first is getting the article into *Physical Review Letters*; the second is getting the boojum into the article. I had been thinking about some puzzling aspects of supercurrents in ^{3}He-A that I had learned about at Erice. When I returned to Ithaca from New Hampshire and continued pondering the problem with some associates, the resolution became apparent. While not an earthshaking advance, it was a likely candidate for one of the major discoveries of the week, and *Physical Review Letters* seemed an entirely appropriate vehicle.

Without any distortion of our central point, I was able to introduce a remark about boojums. Anticipating resistance to the boojums even if the article were accepted, I carefully supplied footnotes documenting the scientific and literary origins of the word. I even added a reference to Webster's second edition. I had no doubt this would be excised by the editor, but I hoped it would convey to him the fact that "boojum" was, after all, no more than an ordinary English common noun, and therefore not a candidate for rigorous editorial scrutiny. Mindful, however, of the mess I would land in if they happened to check my citation with the third edition, I added the phrase

> In this, as in many matters, the views of the 3rd edition should be spurned.

Then I sent the manuscript off.

Surprisingly soon an associate editor of *Physical Review Letters*, Gene Wells, phoned me in my office. Our article had been accepted, he told me. Instantly I readied myself for the central battle of the campaign. *Physical Review Letters* does not announce the acceptance of papers by telephone, unless something is up, and I knew what was up: boojums!

There was a small problem, he explained; in the second paragraph we used a word.... I gave him the essence of my letter to Daunt. To my gratification he acknowledged that this might be a case where a judiciously selected exception could fortify the general rule against neologisms. But the term needed to have a rigorous justification. For example *Physical Review Letters* was even taking a strong stand against "instanton." I congratulated him. But "boojum" was something else. Was it? he wanted to know, and then he put me through a cross-examination such as I have not had since my PhD qualifying exam. What aspect of the Boojum was pertinent? What was it that vanished away? Could the metaphor be construed as mixed? And, perhaps most importantly, if they let me get away with "boojum" would I be back to them with "snark"?

I swore then and there a solemn oath never to try to make the word "snark" an internationally accepted scientific term. I promised never again to try to introduce any new word at all. "Boojum" would be quite enough for me; indeed, the ways things were going it might turn out to be altogether too much.

Wells said they would have to discuss the matter. Some time later he called again. "Boojum" had been approved. But there was the important question of the plural. I had used the form "booja" in our manuscript, not because I favor Latin plurals, but because I had always thought that Boojums was a common name for a rather unpleasantly fluffy kind of beribboned cat. We considered the possibilities. I had already discussed the question that summer in New Hampshire with A. J. Leggett of the University of Sussex, who has made profound contributions to the theory of superfluid helium-3 and who did an undergraduate degree in classics at

Oxford. Leggett's position was simple: It would be evident to any ancient Roman that boojum was a word of foreign origin, and words of foreign origin are indeclinable. Therefore if one did want to form a latinate plural that plural should be not "booja" but "boojum." One boojum, two boojum.

I liked Leggett's logic, but had persisted in using "booja" in the belief that superfluid ^{3}He-A was complicated enough without the added problem of distinguishing singular from plural. Now, however, *Physical Review Letters* and I were setting a standard for the generations to come. I thought (incorrectly, as we shall see) that Leggett's argument against "booja" was unassailable. Wells frowned on "boojum" for the same reasons that I did, and that left us only with "boojums", so "boojums" the plural became.

It was pointed out to me more than two years later at Harvard by Wendell Furry (who, curiously enough, was on the committee that raked me over the coals at my PhD qualifying exam) that the question of the plural is definitively resolved by Carroll himself, the very first time the word appears in the poem:

> "For although common Snarks do no manner of harm,
> Yet I feel it my duty to say,
> Some are Boojums – " The Bellman broke off in alarm,
> For the Baker had fainted away.

I like to think that we arrived at our plural by the same logic that Carroll (a celebrated logician and himself an Oxford man) had followed.

In spite of our concern over the correct plural, when the article appeared a citation in footnote five used the rejected plural form "boojum." I pointed this out in a short note to the editors; soon thereafter the following notice appeared on the Errata page:

> STABILITY OF SUPERFLOW IN ^{3}He-A. P. Bhattacharyya, T.-L. Ho, and N. D. Mermin [*Phys. Rev. Lett.* **39**, 1290 (1977)].
> In the second sentence of the first paragraph following Eq. (7), the symbols K_b and c_0 should be interchanged. The point being made is unaltered by this transposition.
> In Ref. 5, "boojum" should read "boojums".

The debut of "boojum" as a fully authorized scientific term was quiet and dignified:

> ... Surface current can be reduced (with an accompanying reduction of bulk current) by the motion of a special type of surface point singularity (a "boojum").[5,6] The relative importance of these mechanisms depends on details of the pinning, nucleation, and equilibrium populations of boojums, as well as on the energetics of vortex texture formation.

I was surprised to find that my lexicographic footnote had not been deleted:

> [6]N. Webster, in *New International Dictionary* (Merriam, Springfield, Mass., 1934), 2nd edn, p. 308. See also p. 2379. (In this, as in many other matters, the views of the third edition should be spurned.)

Almost a year later I was at a conference on low temperature physics in Grenoble (where I gave a talk, the subsequent publication of which led to the first "boojum" in *Journal de Physique*; the weather was unpleasantly but not remarkably hot and humid). At an outdoor barbecue I happened to meet Gene Wells. After several glasses of wine he confided to me that my lexicographic footnote had won the day for the boojum. It seems that the weight of opinion among the editors was against the term. However the editor-in-chief, George Trigg, had loathed the third edition of Webster's for many years. The unprecedented opportunity I had handed him to print a brisk attack on it in his own journal was more than he could resist; he forsook one set of linguistic principles for a higher one, and let the boojum in.

Having launched the boojum in the grandest style I could manage, I felt my job was done. Whether the ship would stay afloat or softly and suddenly vanish away was in the hands of my fellow physicists. My tension during this long period of waiting was alleviated by an interesting correspondence with Sam Trickey. Trickey edited the proceedings of the Sanibel Symposium. Readers of the article in *Physical Review Letters* were referred to my paper in *Quantum Fluids and Solids*, ed. S. B. Trickey *et al.*, for mathematical details about boojums. I discovered after the article appeared that at least one colleague concluded from the juxtaposition of Trickey and the boojum that the entire thing was a colossal hoax on my part. I convinced him that I was serious, but he was unappeased. He had worked in the office of the Scientific Adviser to the President. "Wait till Proxmire sees *this*!" he warned.

Trickey had a godfatherly interest in the success of the boojum, having allowed me to print "boojum" and "booja" many times in his book. He also entered the word in the index and let me insert among the technical drawings a copy of an illustration of the Baker, being warned by an elderly uncle of the disastrous end that awaits the finder of a boojum. Upon the appearance of the article in *Physical Review Letters* I received a congratulatory note from Trickey who was, however, puzzled by the reference to the dictionary. I had cited the 1934 printing of the second edition; he had looked in the 1930 printing and found nothing at all. "Boojum" was not there. Furthermore "snark" was only listed below the line, where it was dismissed as "VAR of snore."

This discovery set us both off on searches that have yet to conclude. I found a 1940 printing of the second edition that agreed with the 1934 printing except that the risk to hunters of the boojum was that they might "softly and *silently* vanish away." Trickey found a 1951 printing in which it is the boojum itself that softly and suddenly vanishes away. Had boojum fever infected the editors of so staid a publication? Were they enlisting the boojum to help them trap plagiarists? I wish I knew.

While Sam Trickey and I were flipping to the copyright page of many a heavy volume, I again received for review from *Physical Review Letters* a paper of which P. W. Anderson was an author. And it had boojums. The word, in fact, appeared in title, abstract, text, *and* a figure caption. Anderson had just won his Nobel prize. If the paper appeared nothing could sink the boojum. Eagerly I read it, and realized with dismay that it was wrong. I thus faced an unusual moral dilemma.

Relations between authors and referees are, of course, almost always strained. Authors are convinced that the malicious stupidity of the referee is alone preventing them from laying their discoveries before an admiring world. Referees are convinced that authors are too arrogant and obtuse to recognize blatant fallacies in their own reasoning, even when these have been called to their attention with crystalline lucidity. All physicists know this, because all physicists are both authors and referees, but it does no good. The ability of one person to hold both views is an example of what Bohr called complementarity.

In this case, however, the referee wanted the paper to appear more than the authors could have imagined. I nevertheless did the honorable thing. Believing that at best I would be rewarded with invective and abuse and at worst, if I was truly persuasive, I would prevent the culminating moment of my own hard fought campaign, I wrote a long thoughtful report, listing all of my objections.

I received a most courteous reply, thanking me for my help. Many of my suggestions were adopted and many of my objections were deftly and effectively dealt with. The central one was not, though the resubmittal letter politely but firmly insisted that it was.

What to do? The harmony could not survive another exchange of letters. I had been wrong before, particularly with authors associated with Bell Laboratories. And the paper, even though – or, I should say, if – wrong, was undoubtedly thought provoking.

I let it through. And it looked glorious! I display only the grand opening, but there were boojums all the way through:

Boojums in Superfluid ^{3}He-A and Cholesteric Liquid Crystals

D. L. Stein, R. D. Pisarski, and P. W. Anderson
Department of Physics, Joseph Henry Laboratories, Princeton University, Princeton, New Jersey 08540
(Received 26 January 1978)

Because of the similarity of their order parameters, there are close analogies between defects of ^{3}He-A and cholesteric liquid crystals. In particular, boojums, originally predicted

> for ^{3}He-*A*, should exist as well in cholesterics. Certain textures experimentally observed and reported in the literature are identified as boojums. A topological analysis is given, and the effects of boojums on dynamical properties of cholesterics are discussed.

As I was learning, however, things have a curious way of softly and suddenly going awry, when boojums are concerned. I was not to go unpunished for my breach of professional ethics, though retribution was another two years coming.

Meanwhile there were promising developments, of which I mention only two. Here is the first French boojum:

> **Resumé.** – On présente une analyse topologique des configurations de surface et de volume dans les cholestériques et les nématiques. On rappelle et illustre la procédure de Volovik pour combiner défauts de surface et de volume; le problème de la torsion dans cholestériques est discruté suivant le point de vue de Cladis *et al.* Les considérations sont appliquées aux boojums cholestériques. On étudie également la question des textures non singulières, et on propose une texture stable de type soliton pour un cholestérique. Enfin on considère le problème d'un nématique dans une sphère, et on présente les différentes solutions.

It happens to occur in a translation of the English abstract of an English language paper published in a French journal, but for all that it is the kind of clipping one can send to the editor of the *Journal of Low Temperature Physics* with no little satisfaction.

Even better was the first Russian boojum. It occurred in a preprint I received last year from the Landau Institute. *Mermin nazval "budzhum,"* it declares: Mermin called it "boojum." It goes on to explain that the word is taken from *Lyuis Kerrol's "Okhota na Snarka."* The Russian boojum was a counterexample to Leggett's theory that highly inflected languages would treat the word as indeclinable, and demonstrated that *booja* could have been an acceptable plural. In one page I found the nominative plural (*budzhumi*), the genitive plural (*budzhumov*), and, my favorite, the instrumental singular (*budzhumom*).

Fig. 5. Some Russian boojums from the Landau Institute in Moscow. The nominative (*budzhumi*) and instrumental (*budzhumami*) plurals are on display, contrary to the early views of A. J. Leggett.

БУДЖУМЫ В А-ФАЗЕ И ТОПОЛОГИЯ ПОВЕРХНОСТЕЙ

С граничными условиями на поверхности А-фазы связано существование специфических поверхностных особенностей, называемых буджумами.

На границе с сосудом, где вектор **l** зафиксирован, у А-фазы остается лишь одна степень свободы — вращение $\vec{\Delta}'$ и $\vec{\Delta}''$ вокруг **l**. Таким образом, область вырождения А-фазы на поверхности является окружностью.

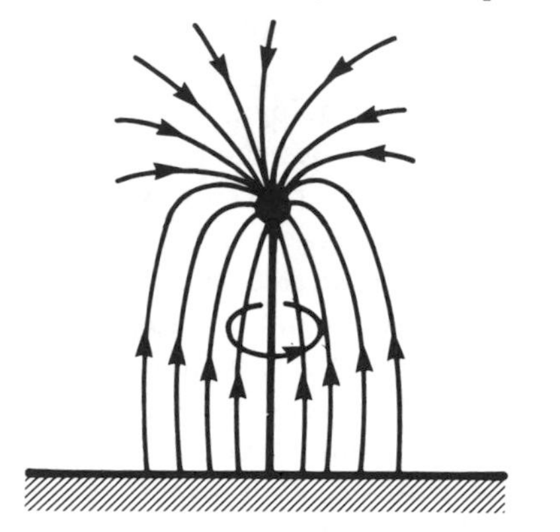

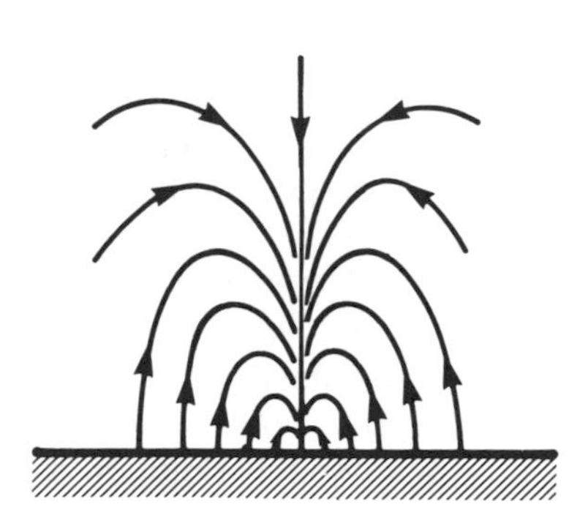

I sent a copy of that page to Leggett. His reply was in the finest tradition of science and linguistics:

> I bow, however reluctantly, to the wisdom of the majority. I anticipate that we shall now presumably be getting, from Accra, reports of M'BOOJUM and from Singapore of BOOJUM-BOOJUM, while those investigated by Olli Lounasmaa will presumably behave BOOJUKSESTI.

On Tuesday, 3 June 1980 my long delayed comeuppance arrived through the unexpected medium of *The New York Times*. An article appeared on whimsy in scientific nomenclature. It talked about quarks for a while, turned to Lewis Carroll, and then finally, at the end, it said

> Some snarks are dangerous to hunt, of course, because they may actually be boojums – beings that annihilate their hunters by making them disappear forever. Boojums found their way into science thanks to Philip W. Anderson, a 1977

> Nobel Prize winner, who needed them as personae in a difficult notion about the broken symmetries of nature.

This terrible thing was brought to my attention, with a smirk, by my own graduate student. "Bell Labs always wins out in the end" he cheerily opined, and danced off.

There was only one thing to do. I wrote a short but firm letter to the *Times*, and gave it to my trusted friend and colleague, N. W. Ashcroft, to send under his own name. He joined the fight with such fine spirit that the version published in the *Times* on 17 June was his own revision of my original letter. I give here the rare but still surviving *urtext*:

> Scientists may be addicted to whimsical nomenclatures as Malcolm Brown suggests (June 3) but in important matters like priority they are deadly serious. Philip W. Anderson did indeed use the term "boojum" to express a difficult notion about the broken symmetries of nature. However N. David Mermin introduced "boojum" for the same purpose over a year before in at least three publications prior to Anderson's venture in boojology. The last of these required Mermin to do extensive battle with the editorial board of one of the world's most distinguished journals of physics, who rightly regarded themselves as guardians of the purity of scientific discourse, and yielded only after Mermin presented a most cogent case that "boojum" was apt. This hard fought test paved the way for Anderson's easy reference to boojums in title, figure caption, and text in that same journal, as well as for the subsequent appearance (just last month) of the first French boojum and Russian budzhum.
>
> Anderson has, as Browne notes, a well deserved Nobel Prize in physics, but Browne's nomenclatural accolades in the quite unrelated matter of boojums are due to Mermin, if truth itself is not to softly and suddenly vanish away.

The day after Ashcroft's letter went off I received a nice note from Anderson:

> Dear David:
>
> I note a depressingly typical example of the Matthew effect in today's *Science Times*. Do you want me to try to correct it? He didn't talk to me or anybody who knew anything.
>
> Regards, and sorry
> *Phil*

I sent back a cheery reply, passing it off as a case of *sic transit gloria boojorum.* I did, however, ask what the Matthew effect was. I got an immediate reply: "Matthew effect: R. Merton: 'to him that hath shall be given, etc.'" I knew R. Merton hadn't said that, and turning to my Bible found that "etc." stood for "but from him that hath not shall be taken away even that which he hath." There I was, once again at the wrong end of the Matthew effect, and not even knowing it until I was told.

But he who deals in boojums does not stay down for long. The *Times* printed Ashcroft's letter; it has yet to print Anderson's reply to Ashcroft's letter (pointing out that while I invented the name, we both independently invented the object). And in that vast land beyond the Bell System where *The New York Times* cannot be had for love or money, they will all shortly be learning that *Mermin nazval budzhum.*

The slavic boojum

The slavic boojum is especially dear to me. I spent the winter of 1978–79 visiting Leggett in Sussex. It was the coldest English winter since 1962–63 (the last winter I had spent in England). In February Leggett and I lectured at a winter school in the little Polish town of Karpacz, in the Sudeten mountains near the Czech border. My friend Ashcroft from Cornell was also a lecturer at the school, and I met him at London airport so we could fly on to Warsaw together. We landed during a lull in the worst Warsaw blizzard of the decade, but the story of our epic journey from Warsaw to Karpacz will have to be told elsewhere. My only point is that the weather was being boojumish again.

So I talked about boojums at the Polish winter school. In the audience were two distinguished physicists from the Landau Institute, where boojums had by then been known and discussed for some time. Because most of the audience knew Russian better than English, when I introduced the boojum I wrote on the blackboard in my crude Cyrillic the first person plural form (*budyem*) of the Russian verb to be, as an aid in pronunciation.

Ashcroft, who was sitting near the two Russians, reported to me later that their heads immediately flew together and from their conference emerged an endless stream of sound: "Boojum, boojum, boojum, boojum, boojum...." Boojum fever, we decided, but on thinking it over I have managed to reconstruct the conversation he overheard. It was completely rational:

Academician K (straining to decipher my handwriting): Budyem?

Academician A (more used to the wretched calligraphy of foreigners): Budyem.

Acad. K (puzzled and surprised): Budyem?!

Acad. A. (explaining why): Budyem – budzhum.

Acad. K (like all Russians, oblivious to any resemblance between the two words): Budyem – budzhum??

Acad. A (confirming this, with more than a touch of disapproval): Budyem – budzhum.

And so on.

I didn't write a word about boojums in the proceedings of the Karpacz Winter School. I didn't need to.

Notes for the General Reader

1 Superfluids are liquids in which currents (*supercurrents*) can flow forever, without succumbing to the frictional drag that causes currents in ordinary fluids to die away.
2 An anisotropic liquid is one whose atomic structure in any little region points along a particular line.
3 A vortex line is the long funnel of a little whirlpool.

Postscript

I am pleased to announce that in February, 1988, "boojum" appeared in the instrumental plural (*budzhumami*) in the *title* of an article in the Soviet version of *Physical Review Letters*. See O. D. Lavrentovich and S. S. Rozhkov, *Pisma v ZhETF* **47**, 210 (1988).

2

Commencement address, St. John's College, Santa Fe, May 18, 1986

I have to say that the only other time I was asked to talk at a commencement was 1952, when I graduated from high school. So while I suppose I should have spent the last month thinking hard about the great challenges lying ahead for all of you, I was actually more preoccupied with the great challenge lying ahead for me. What can a middle aged theoretical physicist have to say to the graduating class of this unique college?

The answer came to me a few weeks ago, when I read in a pamphlet about St. John's College that the principal goal of a liberal education is to acquire the skills of *rational thought*, *careful analysis*, *logical choice*, *imaginative experimentation*, and *clear communication*. Having always regarded these as the primary tools of the physicist, I realized that I could do no better than to call to your attention a few examples of the application of these skills in public affairs, in private life, and on the frontiers of science.

Let's begin with *clear communication* in public affairs. Several years ago I was half listening to an early speech by a new President who was acquiring a reputation as a clear communicator. Talking about a trillion dollar national debt, he was saying: "A trillion dollars is so much money that it's hard to grasp the idea, so I want to tell you how to make it a little more real."

Instantly, the President had my full attention. I spend a lot of time trying to get students to think meaningfully about quantitative information, and I hate the mindlessness with which people publicly discuss numerical facts without any attention to scale. Here, I thought was a new moment in public discourse. This President – this Clear Communicator – was probably going to explain that there were a quarter of a billion people in the country, so a trillion dollars was $4000 per person, or $16 000 per family of four. He would then go on to compare this public debt with the personal debt of such a family – maybe $40 000 on a home mortgage and $8000 on an automobile, and he would then discuss whether this 1 to 3 ratio of public to personal debt was or was not reasonable. A new era of *rational thought* was about to dawn.

That's what I expected. What he actually said was something like this: "If you took a trillion dollars in one dollar bills and stacked them on top of each other, the pile would reach halfway to the moon."

I had two reactions to this, and I maintain that the liberally educated person should have both. First, and most importantly, in the matter of *clear communication*, disappointment and dismay at this triumphant substitution of one meaningless number for another. Second, an annoying but irresistible urge – the unbreakable habit of one trained in *careful analysis* – to check this particular piece of foolishness. Was the President right?

Well, a dollar is a pretty sturdy piece of paper. A book of well made pages like that would probably be an inch thick if it had 400 pages. Don't forget that 400 pages are only 200 pieces of paper, so an inch is $200. A foot (10 inches) is $2000. A mile (5000 feet) is $10 million, a hundred miles is $1 billion, and a hundred thousand miles – half way to the moon – is $1 trillion. Right on! God protect us from such misuses of the noble art of arithmetic, but if you've got to do it, at least do it right.

Life will continually present you with such technically correct but wacky and wildly wrongheaded appeals to the fruits of arithmetic. Watch out for them.

Equally abundant are similar invocations of science. Last March I was at a meeting of the American Physical Society, which, improbably, took place in Las Vegas – the real Las Vegas – among the slot machines and blackjack tables of a huge establishment called the MGM Grand Hotel. After two days in this genuinely lunatic environment I couldn't stand it anymore. I got a rental car, and drove out to Hoover dam, which I'd always wanted to see. The contrast was powerful and dramatic between the dignity of this immense and overwhelmingly purposeful piece of architecture, and the vast monuments to bad taste and pointless frenzied activity that I had just escaped from, but what particularly intrigued me was a great plaza on the Nevada side of the dam built to commemorate the formal dedication by President Roosevelt.

Set in gleaming brass in an enormous marble pavement flanked at either end by two fierce, lean, gigantic white angels, was – what do you think? Surely a map of the Colorado river and the surrounding lands that this wonderful structure would fertilize and protect? No – not at all. Imbedded in that vast stretch of marble were a great many circular brass disks. They represented the positions of the planets with respect to the fixed stars on that day, September 30, 1935 that President Roosevelt dedicated the dam.

The attention to detail was impressive – the disks came in half a dozen sizes, representing stars of different magnitudes. Most remarkably, embedded in the pavement, in brass letters almost an inch high, was a giant astronomy lesson. To read it, you had to walk back and forth along each line of text, squinting in the blinding sunlight. But if you worked hard, walking back and forth, you could learn about red giants, white dwarves, cepheid variables, red shifts, and the expanding universe.

What did any of this have to do with Hoover dam? Absolutely nothing. It was an expression of exuberance: "We've conquered the Colorado river, and we've frozen the planets in place against the stars to help any visitors from outer space figure out when we did it. And please note that we know one hell of a lot about those stars."

Is Hoover dam a contribution to science? Certainly not. It is a contribution to human welfare, a spectacular achievement of engineering, and a great work of art. Somehow that wasn't enough, and it had to be additionally recorded in brass and marble that the achievement resembled the discovery of the expanding universe.

The world will continually present you with opportunities like these, to misinterpret big numbers and confuse with science things that are not. Often critical questions of *logical choice* depend on making the right interpretation or avoiding the confusion.

Fifty years after the construction of Hoover dam, the great public works projects of our time are done not by the Army Corps of Engineers, but by the National Aeronautics and Space Administration. A new space shuttle will cost a couple of billion dollars to build. How does one find the measure of such a sum? Not in a stack of ones stretching from Albuquerque to Los Alamos. Nor as the yearly income of 30 000 commencement speakers, nor even as a tenth of one percent of the national debt.

No. A couple of billion dollars is somewhat larger than the annual budget of the entire National Science Foundation, the nation's most important source of support for pure research in all the sciences. Why do we need another space shuttle? For the science. What science? Well, maybe it will be possible to make more perfect ball bearings in a weightless environment, or grow more perfect crystals, or improve the manufacture of pharmaceuticals. Ah, then it must be being done as part of the crash national program in crystal growing, pharmaceutical, and ball bearing betterment. No. It's being done for many complex reasons, but few of them have anything to do with science.

The space shuttle teaches us that *careful analysis* and *rational thought* have suffered a decline since 1935. In 1935 we needed a dam, built it, and then, in a ceremonial moment, tried to lend it an absurdly inappropriate cosmic significance. Today we build on that grand scale out of some inarticulated sense of cosmic significance, and then search around frantically for a purpose.

Watch out also for heroic efforts, launched on behalf of an articulated but manifestly preposterous goal. Recently, scientists have been called upon to render nuclear weapons "impotent and obsolete." This is a new level of confusion – the confusion of science with magic. The level of expenditure is also new: twenty-five billion dollars over five years just for research and development.

Scientists and engineers are almost unanimous in agreeing that the protection of people and cities that this program ostensibly aims at is impossible. Powerful demonstrations of its absurdity can be found in the pages of *Scientific American* or even the *New York Times*. Over half the faculty of the top twenty physics departments in the nation have signed a pledge not to participate in such folly. Even the organization established to carry out this parody of a research program acknowledges, when pressed, that the announced goal is impossible to achieve. Yet it looks as if spending at the rate of five billion dollars a year will be approved for a fantasy that will, inevitably, drain the resources for real scientific investigation, which pays some attention to where we are and where we can get to from there, as well as where we might like to be.

So much for rational thought in public affairs. President Delattre told me that a good commencement address should illustrate how intelligence and the liberal arts can affect a life, so I also tried to think on a very small scale, of characteristic incidents in my own life. I offer you two:

The first I flatter myself comes under the category of *imaginative experimentation.* I've lived for several years in England, a country where people drive on the wrong side of the road. I discovered that in England I could not correctly use the words right and left when I was in a car. Something deep within me knows ineradicably that "right" means "easy turn" and "left" means "hard turn"; and I invariably produced the wrong term when I tried to give directions.

More recently I was in Japan where they also drive on the wrong side, and I had to learn enough Japanese to give directions to cab drivers. Double trouble. Getting straight the difference between right and left is hard enough in a language where each is represented by an entirely meaningless sound. Now added to that difficulty, was the further problem of interchanging them while driving along the wrong side of the road.

The solution came to me in a great flash, and the same sense of pride and joy that accompanies a discovery in physics: don't learn the terms as "left" and "right" – learn them as "easy turn" and "hard turn." "*Hidari e magatte kudasai*" – "Please make the easy turn." It worked perfectly, and I see no flaw in my solution, until the improbable day I find myself giving instructions in Japanese in a country where they drive on the right.

Here's a second example of how *careful analysis* and *imaginative experimentation* can be put to work in your own lives. Suppose you add up the results of many measurements. Each has a certain error. How big is the error in the total? Well, you might think if you were adding up N items, the error in the total would just be N times the error in each. But it's not that bad. Some of the measurements will be too big, some too little, and when you add them up the errors start to cancel each other. In fact it's a famous theorem (which is very easy to prove) (and I hope you learned it in your four years here) that the error in the total is not bigger than the individual error by N, the number of measurements, but only by the square root of N, a much smaller number. If you're adding 100 items, the error in the total will not be a hundred, but only ten times the individual error.

You can put this principle to good use relieving tedium at the supermarket and checking the accuracy of the clerk. As you put items or groups of small items on the counter it's very easy to keep a running total of the price in your head if you estimate each item only to the nearest dollar – hardly more difficult than just counting what you put down. The size of the typical error in each estimate is 25 cents. If you have 35 items, then the square root of 35 is about 6 and so the typical error in your total will be 6×25 cents or about a dollar and a half. With a little practice you can begin to develop a sense of whether you've been rounding up more than you've been rounding down, and make the occasional adjustment to improve your total. I'm at the point where I often get to within a dollar of the right answer. My children are enormously impressed (or embarrassed) when I announce the total ("looks like about $43") just before the clerk rings up $42.38.

What is the point of all these examples? What have they to do with the life of a scientist, or the attitudes of the liberally educated? They share a few principles of almost banal simplicity: one should take nothing for granted; one should try to understand everything; one should constantly look for new ways to deal with old ideas, new ways to apply knowledge. One should take delight in surprises – in turning things upside down. In discovering that one was entirely wrong about something obvious.

One of the classic examples of this in the 20th century is the theory of relativity. You remember Newton's wonderful statement at the beginning of the *Principia*: "Absolute, true, and mathematical time, of itself, and from its own nature, flows equably without relation to anything external...." Lovely as this is, as a description of nature one might think it was trivially self evident, or one might think it was devoid of content. Neither view is correct: as you know, it is simply wrong.

There is no absolute, true, and mathematical time. One man's now is another man's then is a third man's yet to be. Time is something we impose on nature, as we impose lines of latitude and longitude on the world. Minkowski's poem isn't as magnificent as Newton's, but it has the advantage of truth: "Henceforth space by itself, and time by itself, are doomed to fade away into mere shadows, and only a kind of union of the two will preserve an independent reality."

This is a good thing to know. If this astonishing discovery had not been made we would not today have anything remotely comparable to our present understanding of matter or the universe. But, just as importantly, in making this discovery or learning about it, we discover something very important about ourselves: that everything, no matter how evident or obvious, should be doubted, questioned, viewed with suspicion; that unexamined truths are likely to be falsehoods; and that there is much to be gained from the discovery that one has been deeply, persistently, and utterly wrong.

Here is a final example from my own work as a physicist in the last year and a half. Matter likes to solidify into crystals. The characteristic feature of crystals is that their atoms are arranged in regular periodic patterns, that repeat themselves like the design on a tile floor. Such patterns are often quite symmetric: if you take a crystal and rotate it appropriately it ends up looking exactly the same as before. It's very easy to prove that not any rotation is consistent with the periodicity of the crystal. Crystals can be symmetric under rotations that are a half, a third, a quarter, or a sixth of a complete revolution, but nothing else. (This is another nice theorem – if I had two hours and a blackboard I could teach you a lot of nice things.) In particular they cannot be symmetric under five-fold rotations, that are one-fifth of a complete revolution.

This means that of the Platonic solids only the tetrahedron, octahedron, and cube can have the same rotational symmetries as a crystal. The icosahedron and dodecahedron have five-fold rotational symmetries, and these are impossible.

This has been known for as long as the atomic theory of matter has been accepted, and has been one of the pillars of crystallographic science. It therefore caused a general sensation, when about a year and a half ago an alloy was produced whose electron diffraction pattern had certain characteristic features that occur only in crystalline matter, but with a rotational symmetry precisely that of the icosahedron (or dodecahedron). To get a sense of how revolutionary this is, you should know that Landau and Lifshitz, probably the most authoritative series of physics texts in the world, baldly states that icosahedral symmetry is "of no physical interest," since it does not occur in Nature as a symmetry of inorganic matter.

This is the context in which it is the most fun to do physics. Everything has to be reexamined. We thought we knew all the ways atoms might arrange themselves into chunks of matter, but now we've found stuff that has some properties we were sure only crystals could have, and another property we know no crystals can have. All the old ideas have to be viewed with suspicion – all the obvious old definitions have to be reexamined. Any idea is worth trying out, and most of them will probably have to be rejected. Papers are flying back and forth – new experiments are being done all over the world – so many conferences are being organized that if you went to half of them you wouldn't have any time to think about the problem at all. Priority disputes have erupted. The crystallographers think the physicists have gone crazy. The physicists deplore the limited imagination of the crystallographers. It's science at its raucous best.

This is not how you go about building Hoover dam. This stuff was found by accident. We have no idea what it's good for or where it will lead us. The entire national effort in this field probably amounts to a few hundred thousand dollars – to be viewed not as a hundred foot stack of ones, but as the cost of supporting a few dozen graduate students, since most of the experiments are easily done with existing equipment. These icosahedral quasicrystals are providing us a most excellent occasion for exercising the skills of rational thought, careful analysis, logical choice, imaginative experimentation, and clear communication.

The best I can wish for you on this commencement day is that you too will continue to find many such occasions.

3

"One of the great physicists... and great characters"

Review of *Landau: a great physicist and teacher*
By Anna Livanova,
Pergamon, Elmsford, N.Y., 1980.

Imagine a *New Yorker* profile in the literary style of *Krokodil*, and you have this book. Not a biography of Lev Davidovich Landau, it is an attempt to portray the human and scientific style of this extraordinary man and the remarkable school of theoretical physics he single-handedly created and sustained. Written for the edification of youth, it suffers from that turgid sentimentality – the rapture of an elephant in heat – that seems to accompany all Soviet efforts at morally elevating prose. It should nevertheless be fascinating reading for anyone working along the trails first blazed by Landau himself, that is, for almost all physicists.

After two introductions (why two? why not, I suppose, but this is typical of the literary anarchy that follows) Anna Livanova gives us a tantalizing glimpse of Landau before the age of sixteen, when he entered the university in Leningrad. Aside from a nod to his wife (an engineer in a chocolate factory; love at first sight; a son, Igor, in 1946) the rest is, so to speak, history, and the history is told (and this is the saving grace of the book) primarily through a wonderful series of anecdotes and vignettes.

There are two main sections: an evocation of the school of Landau as it operated under the Master, and an attempt to convey in nontechnical terms the nature of his contributions to the theory of superfluid helium. The former is the better part. The weekly theoretical seminar is brought vividly to life: "Noise, shouts, questions continually interrupting and annoying the speaker, exchange of tart repartee, no hesitation in calling out 'nonsense,' 'idiotic,' 'crazy,' 'pathology,' and even more scathing comments.... Cossack anarchy." He presided over it like a conductor. M. I. Kaganov's assessment is all the more powerful as an island in the sea of inspirational hyperbole: "Landau had an impressive combination of quick reactions with informed knowledge and profound understanding. I have never seen anyone like him." Entrance to the school was through the "theoretical minimum," Landau's *Gradus ad Parnassum*. Anyone could take it, but was mastered by only 43 people between 1933 and 1961, all examined (though, toward the end, only in part) by Landau himself, the names of the survivors being duly inscribed by him in a book.

I was interested to learn that Landau suffered from writer's block. It is well known that Evgenii Lifshitz single-handedly cared for the literary side of the *Course of Theoretical Physics*, but "even the papers containing only his work, without co-workers, were written for him by Lifshitz, from the middle 1930s until the end." In an Appendix (one of the high points of the book, together with a marvelous preface by Rudolf Peierls) Lifshitz tells us that Landau's "insuperable desire for brevity and clarity of expression forced him to devote so much time to the choice of each phrase, that ultimately the task of writing anything, be it a scientific article or a personal letter, became a torment." Could this be the origin of that tight and terse style that characterizes to this day the writings of the Soviet school?

It is the many views of the man and his way of working that give the book its real power. Aspects of the story that Soviet youth might find unedifying, however, receive a curious treatment. A single remark hints at the extent of Landau's celebrated problems with the Stalin regime: "In one very burdensome year Landau reconstructed for himself the theory of shock waves. He made all the calculations mentally, without paper and pencil." Also in the words-of-one-syllable department is the account of the termination of Petr Leonidovich Kapitza's Cambridge career: "Arriving on holiday in the summer of 1934, he learned that he must now work in the Soviet Union." Kafka could not have put it better.

In spite of – perhaps, in some ways, because of – its limitations, this is an intriguing and strangely moving book. It abounds with gems: a short appearance by one Genia Kannegiesser, the belle of Leningrad physics, now Lady Peierls; Landau's 50th birthday party ("Everybody's fifty now, except the kids"); Landau's classification of conversations, and of physicists (the latter on a scale of 1 to 5, Einsten rating $\frac{1}{2}$, Bohr and the other founding fathers 1, Landau himself, $2\frac{1}{2}$ (but promoted later in life to 2), and the 5s – "pathologists"); in keeping with this scheme, Landau, in thoughtful reply to a young lady: "No, I'm not a genius. Bohr is, and Einstein is. I'm not. But I'm very talented.... Yes, I'm very talented."

One finishes the book yearning for a real biography. If people are not already at work on one, someone should begin while Landau is still so vibrantly alive in the memories of his friends and pupils, those fortunate enough to have known (in Peierls' words) "one of the great physicists and one of the great characters of our generation."

4

My life with Landau

Talk delivered at the Landau Memorial Conference,
Tel Aviv, June, 1988.

Last October I received a phone call asking if I could survey Landau's contributions to condensed matter physics at a commemorative 80th birthday symposium to be held in Tel Aviv the following June. I said no, I couldn't. I wasn't an historian of science, and wouldn't pretend to be one. And as a physicist I felt that Landau's contributions in condensed matter physics spanned an area broader than my competence to review. Then I hesitated. I never had the opportunity to meet Landau, but there have been times in my life when I have felt his intellectual presence so vividly that, more than a teacher, he appeared to me almost as a scientific muse. How could I turn down an opportunity to pay public homage to the man who, more than any other, built and permeates the edifice, poking around the corners of which has been the largest part of my experience as a physicist?

So I told my caller that although I couldn't talk on the topic he wanted, I could talk about the ways in which my own career has intertwined with the thoughts of Landau; how no matter what new area I turned my attention to, I invariably found myself working with concepts of his devising. Such a talk would not be historical. It would simply illustrate how the work of Landau impinged on a typical run-of-the-mill condensed matter physicist of a later generation, and it might more appropriately be titled "My life with Landau." "Great," said my caller, "just what we want." And so I received a month or two later, rather to my alarm, an announcement telling me that I, who had never met the man, would be giving a talk in Tel Aviv on my life with Landau.

Three weeks ago I started thinking about the trouble I had got myself into. The first thing I realized was that although I had indeed made it clear that I was not an historian of physics, I was dangerously close to embarking on a new field, that might be called the chutzpahlogy of physics. For by undertaking to examine the career of Landau from the perspective of my own, I could hardly avoid telling you as much, if not more, about myself than about the person you are interested in hearing about. But that was the agreement. Therefore I began by examining Landau's publication list – exactly 100 entries, as listed in ter Haar's edition of the Collected Papers, and then I asked my secretary to run off an updated version of my own publication list. I found, disconcertingly, that it too had exactly 100 entries. True, Landau's list, most unfortunately, has ceased to grow while mine, some might argue also unfortunately, has not. But when I next noticed that Landau was 53 when he wrote his last paper and I am now also 53, I decided that God and Landau must have got together and decided that if anybody could pull off an acceptable act of chutzpahlogy, it was me. So here I go.

Let me first modify my title. There is a short biography of Landau by Anna Livanova. There she describes his classification of physicists on a logarithmic scale from 1 to 5. Einstein has a special rank, $\frac{1}{2}$. Bohr, Schrödinger, Heisenberg, and a few others were 1s. Landau, originally a $2\frac{1}{2}$, later promoted himself to a 2. Members of class 5 were called "pathologists." When I reviewed[1] this book for *Physics Today* the editor asked for a short biographical blurb. Just say, I replied, that I've worked in fields of interest to Landau in the hope of getting out of class 5. So, in the June 1981 *Physics Today*, you can read at the bottom of page 61, "N. D. Mermin is professor of physics at Cornell University. Struggling to qualify as a $4\frac{1}{2}$, he has made occasional contributions to the theory of phase transitions and liquid helium." My modified title is "My life with Landau: homage of a $4\frac{1}{2}$ to a 2."

[1] Reproduced here in the previous chapter.

I did my PhD research thirty years ago at Harvard. It was then the practice at Harvard never to mention the names of non-Harvard physicists, so I knew of Landau only as the coauthor of a formidable series of textbooks, until I stumbled across him in the middle of my thesis. My adviser had asked me to investigate whether an instability that arose in a certain crude calculation of the two particle propagator in the particle–hole channel might not herald the onset of a different kind of superconductivity from the type that Bardeen, Cooper, and Schrieffer had recently demonstrated was associated with an instability in the particle–particle channel. (The result of my investigation, by the way, was no, it did not.) In the course of contemplating density autocorrelation functions in the complex frequency plane, I discovered that there were poles other than the one preoccupying me, associated with damped rather than uncontrollably growing modes. Another student with an interest in plasma physics told me that these described something called "Landau damping."

So I first met Landau as a plasma physicist and fellow explorer of the complex frequency plane. In preparing this talk I looked up the "Landau damping" paper ("On the vibrations of the electronic plasma," 1946) in which he gives a general solution of the collisionless Boltzmann equation for a one-component plasma – now commonly called the Vlasov equation – paying careful attention, as Vlasov evidently did not, to the initial conditions which determine unambiguously how the pole in the response function is to be treated. Landau's characteristic style is clearly on display. "These equations," he says on the first page, "were used by A. A. Vlasov for an investigation of the vibrations of plasma. However, most of his results turn out to be incorrect." I remark in passing that it is a pity that such literary directness died with Landau; the journals might be far less cluttered with the imperfect efforts of 5s, 4s, and even 3s, if we ran a significant risk of this kind of public rebuke.

Although I failed to find superconductivity in the complex plane, I did return to the real axis with a PhD, and went off to a Varenna summer school to broaden my horizons, both scientific and geographic. The school was on liquid helium, and everybody kept mentioning this plasma physicist Landau again. He was all over ^{4}He with his two fluid hydrodynamics, his phonons and his rotons. And then when the subject changed to ^{3}He there he was again, with his Fermi liquid theory and a prediction about some funny thing called zero sound.

I have to confess that on my first exposure I entirely missed what is so wonderful about this stuff. We $4\frac{1}{2}$s can be a little slow to catch on. The theory of ^{4}He came through to me as a somewhat *ad hoc* phenomenology, and the Fermi liquid theory seemed to me a modest generalization of Hartree–Fock theory. I did not grasp Landau's characteristic power, austerity, and physical rigor, until I had to teach the subject myself several years later. Take the two fluid hydrodynamics. Once one has introduced the two velocity fields and grasped their general character (the hard part, of course), then their entire hydrodynamics (in Landau's own words (1941)) "can be obtained absolutely unambiguously starting from the one condition only that they should satisfy all the conservation laws." I would guess that the power of this approach was not fully appreciated outside of Landau's own school for nearly two decades.

Then there is the concept of the elementary excitation. I quote from Landau's 1949 polemic against Tizsa: "It follows unambiguously from quantum mechanics that for every slightly excited macroscopic system a conception can be introduced of 'elementary excitations,' which describe the 'collective' motion of the particles and which have certain energies ε and momenta p. . . . It is this assumption, indisputable in my opinion, which is the basis of the microscopical part of my theory." Tisza's failure to grasp the character of Landau's approach clearly occasioned a certain irritation in Landau, who remarks, for example, in an opening footnote: "I am glad to use this occasion to pay tribute to L. Tisza for introducing, as early as 1938, the conception of the macroscopical description of helium II by dividing its density into two parts and introducing, correspondingly, two velocity fields. This made it possible for him to predict two kinds of sound waves in helium II.... However his entire quantitative theory (microscopic as well as thermodynamic–hydrodynamic) is in my opinion, entirely incorrect."

An examination of the form of the momentum density in helium-II already affords an illustration of Landau's insight and characteristic style. One writes (and Tisza, apparently unbeknownst to Landau, had written) that the momentum density in a superfluid has the form

$$\mathbf{g} = \rho_s \mathbf{v}_s + \rho_n \mathbf{v}_n \tag{1}$$

where the total mass density of the entire fluid is

$$\rho = \rho_s + \rho_n. \tag{2}$$

The naive view of these relations is that the total mass of the fluid consists of two components: a superfluid component of mass density ρ_s and a normal fluid of mass density ρ_n, and each component moves independently with its own characteristic velocity. This point of view is fine, provided you know what it really means. If you don't know then you are tempted (incorrectly) to view helium-II as being like an ordinary two component fluid (which it isn't) and furthermore, it is not clear how, for example, or why the mass density is to be divided up in this way.

The Landau point of view is quite different. One writes the momentum density as

$$\mathbf{g} = \rho \mathbf{v}_s + \rho_n(\mathbf{v}_n - \mathbf{v}_s) \tag{3}$$

which is the same as the other form provided ρ_s is defined by the equation for the total mass density. But the meaning is now very different. The first term (in which the coefficient is the entire mass density) is to be viewed as the momentum associated with the entire fluid when it is in its ground state in a frame moving with velocity $\mathbf{v}_s$. This moving ground state can be viewed as a kind of background or ether, providing the stage on which the excitations perform. The second term gives the further contribution to the momentum from the elementary excitations which are present at non-zero temperature. In a hydrodynamic theory (and the two fluid model is a hydrodynamic theory) the excitations will be in local equilibrium in a frame with velocity $\mathbf{v}_n$ not necessarily equal to the velocity $\mathbf{v}_s$ characteristic of the background, and their contribution to the momentum will be proportional to the relative velocity of these two frames, whence the second term in $\mathbf{g}$.

Not only does this more sophisticated point of view tell you what the "two fluids" are, but it also provides a quantitative expression for the contribution of the excitations to the momentum, and hence for ρ_n:

$$\rho_n(\mathbf{v}_n - \mathbf{v}_s) = \int \mathrm{d}\tau \mathbf{p} n_{\mathrm{eq}}(\varepsilon + \mathbf{p}\cdot\mathbf{v}_s - \mathbf{p}\cdot\mathbf{v}_n). \tag{4}$$

The argument of the equilibrium distribution function has the following interpretation: $\varepsilon(\mathbf{p})$ gives the energy–momentum relation for the excitations in the frame in which the background is at rest; $\mathbf{p}\cdot\mathbf{v}_s$ corrects this by the Galilean transformation to the frame in which the background moves with velocity $\mathbf{v}_s$; and $-\mathbf{p}\cdot\mathbf{v}_n$ is there because the local equilibrium distribution n_{eq} must be adjusted to give local equilibrium in a frame of reference moving with velocity $\mathbf{v}_n$. In this way one relates the density of the normal fluid – a hydrodynamic parameter – to the microscopic theory of the low lying excited states of helium-II.

As for ^{3}He, the theory of zero sound in a Fermi liquid, far from being the minor modification of Hartree–Fock theory I took it for on my first exposure, is a remarkably general extraction of the observable consequences of a quantum liquid possessing a Fermi spectrum of elementary excitations (of which Hartree–Fock theory is a very special limiting case). As noted, it took me several years to appreciate these points, and since this talk is about my life with Landau, we will return to them at the appropriate stage of my development. Chutzpahlogy!

Meanwhile, I went from Varenna to postdocs in Birmingham in the last of the glorious years when Peierls presided there, and then to La Jolla with Walter Kohn. During that period I became interested in the possibility of exotic quantum effects in metals, semimetals, or semiconductors resulting from orbit quantization in strong magnetic fields. I will not tell you how I failed to find the quantized Hall effect (why should a practical hard headed fellow be interested in applying his results to two dimensions?) The pertinent point is only that there was this guy again – this plasma physicist with a strong interest in liquid helium – the same guy! This time the relevant one particle states were actually named after him – Landau levels. So was their contribution to the susceptibility – Landau diamagnetism. I couldn't get away from him!

Onc of the pleasures of this exercise in chutzpahlogy has been the excuse it has given me to read or reread many of Landau's great papers. In his 1930 calculation of the Landau diamagnetism he comments on an annoying mathematical subtlety that many people would have overlooked: "at very low temperatures and in strong fields... [there should be] a complicated, no longer linear dependence of the magnetic moment on *H*, which should have a very strong periodicity in the field." He remarks, however, that even weak inhomogeneities in the field should wash out this contribution, so "it should be hardly possible to observe this phenomenon experimentally." The experimental observation of this phenomenon was, of course, reported by de Haas and van Alphen in the very same year. Subsequently the de Hass–van Alphen effect turned out to be the single most important probe of electronic structure in metals. I do not know when Landau first learned of their discovery, but he did write a short Appendix to a later paper of Schönberg, giving a neat way to calculate the oscillations.

From La Jolla I went to Cornell, where I have been ever since. I spent my first year there mainly talking to John Wilkins and Vinay Ambegaokar about superconductivity; I learned from them that a principal conceptual tool was something called Landau–Ginzburg theory – I was beginning to get used to running into him by now.

A few weeks ago I reexamined Landau and Ginzburg's 1950 paper, to remind myself what they had to say about their remarkable complex order parameter Ψ, seven years before BCS theory. (Please note in passing that Landau invented the notion of order parameter – Landau and Ginzburg refer their reader to the

1940 edition of the *Statistical Physics* (Landau and Lifshitz).) They say: "We shall start from the idea that Ψ represents some 'effective' wave-function of the 'superconducting electrons.'" There are quotes around "effective" and "superconducting electrons" which honestly acknowledge what was missing in 1950, and I had always assumed that this was the extent of Landau and Ginzburg's intuition. But to my surprise, a few pages further on, they explicitly state that the density matrix $\rho(\mathbf{r}, \mathbf{r}')$ has the form $\Psi^*(\mathbf{r})\Psi(\mathbf{r}')$ as $\mathbf{r}$ and $\mathbf{r}'$ get far apart. True, it is the one- and not the two-particle density matrix from which Ψ is thus extracted (and "e is a charge, which there is no reason to consider as different from the electronic charge.") But aside from that here is off diagonal long range order, twelve years before it was touted by Yang, seven years before Onsager and Penrose, and a year before it was mentioned in a little known paper by Penrose (which I only came upon when trying to hunt down the origin of the concept while putting together this essay).

I was surprised to find in the Landau–Ginzburg paper this correct association of the order parameter with the long-range part of a quantum mechanical density matrix, because it is so conspicuously absent from all of Landau's writings on helium. Such an identification would have led immediately to the superfluid velocity as the gradient of a phase, and hence to $\nabla \times \mathbf{v}_s = 0$, which Landau certainly gets in 1941, but by an un-Landauesque and not very convincing exercise in the formal commutation relations of current and density fields and without, of course, quantization of circulation. But such an identification would also have led immediately to the identification of the lambda transition with Bose–Einstein condensation, a notion that Landau seemed strangely unwilling to accept. Some historian of physics should really try to get to the bottom of this, while contemporaries of Landau are still with us.

Being now a professor, I discovered the deep truth that you learn a subject not by taking a course in it, but only by teaching it. I taught a course in field theoretic methods in condensed matter physics, a subject that was being enthusiastically developed at

Harvard during my student days. The first thing I learned was that this methodology had been not only developed but powerfully applied several years earlier by – surprise – Landau and his school. (As I remarked earlier, in those days the non-Harvard world really had to hit you over the head to get your attention.) Fermi liquid theory was one of the phenomenologies to which these more powerful techniques could be applied. It was only when teaching it that I grasped its bold and powerful logic. Suppose we try to approximate the energies of the low lying excited states of a strongly interacting system by a Fermi spectrum – i.e. we take them to be characterized by a set of quantum numbers that are 0 or 1, with energies that are linear in those quantum numbers: $E = \sum n_j \varepsilon_j$. Then one can view the states as describing a set of quasiholes with $\varepsilon < \varepsilon_F$ and quasiparticles with $\varepsilon > \varepsilon_F$. The lowest states only have such excitations near ε_F. Elementary phase space arguments show that a state consisting of a quasihole or quasiparticle near the Fermi level is negligibly perturbed by any residual interactions. The possibility therefore emerges that the entire picture of weakly interacting quasiholes and quasiparticles can be made self consistent, for phenomena, such as hydrodynamic phenomena, in which only low lying excited states play a role – a sort of low energy asymptotic freedom.

One then extends the picture from excitations around the ground state to excitations about states with slightly distorted Fermi distributions. These are precisely the states that enter into a low temperature transport theory, and the only modification they require is that the energies of the still asymptotically free quasiparticles can depend on the form of the non-equilibrium distribution, through Landau's "f-function":

$$\delta\varepsilon(\mathbf{p}) = \int \mathrm{d}\tau' f(\mathbf{p}, \mathbf{p}')\delta n(\mathbf{p}'). \tag{5}$$

One then plays some magic tricks, (a) viewing the ground state at a slightly higher density as a distortion of the ground state at the original density and (b) viewing the ground state in a moving frame as a distortion of the ground state in the lab frame. This gives important information about the *f*-function. Furthermore – as a bonus at the end – the whole simple phenomenological scheme can be confirmed to all orders in perturbation theory provided the single particle propagator has the appropriate analytic structure.

All this is applied to an asymptotically exact transport theory (low temperature, weak disturbances) and one discovers the generic possibility of non-hydrodynamic oscillatory modes of the liquid – zero sound. Ten years later zero sound was definitively observed in liquid ^{3}He. Twenty years later it was used as the major probe for studying the newly discovered superfluid states of that liquid.

One of the high points of my life with Landau came when I studied his field theoretic derivation of the transport equation for a Fermi liquid and realized that a constraint on the quasiparticle scattering amplitude implied by the exclusion principle could be made the basis for a proof that zero sound was not only a generic possibility in a "normal Fermi liquid," but a necessity. I published this result, describing it as a scholarly contribution to an arcane body of lore, but to my great delight the Landau school noticed it. (Nobody else did). It is now enshrined in a footnote on p. 71 of Part 2 of Lifshitz and Pitaevskii's update of the *Statistical Physics*. Landau is the subject of the opposite footnote on p. 70. Who could ask for a greater reward?

While lecturing on Landau's Fermi Liquid Theory in the Canadian Rockies the next summer, I learned of a field theoretic argument by Pierre Hohenberg that Bose–Einstein condensation could not occur in one or two dimensions. The argument appeared to be rigorous – at least in the sense that physicists customarily use that term – and in a subsequent attempt to convince Michael Fisher of this it occurred to me that the argument could be presented on a level that would make it rigorous even in his sense of the word if it was translated into a proof that there could be no ferromagnetism in a one- or two-dimensional isotropic Heisenberg model. Herbert Wagner and I figured out how to do this.

Having developed some facility with the technique, I next applied Hohenberg's trick to proving more generally Peierls' observation that at least in the harmonic approximation there could be no crystalline ordering in a one- or two-dimensional solid. The Hohenberg argument was based on an inequality of Bogoliubov, whose Moscow school of physics was absolutely orthogonal to the school of Landau. Falling into bad company, I thought to myself. But then I discovered that the Peierls conclusion had been reached independently by Landau, phenomenologically but also anharmonically, at about the same time in 1937.

This development was, in fact, the occasion for perhaps the low point of my life with Landau. I had remarked to Fisher that the argument could be viewed not as ruling out the existence of two-dimensional crystals, but merely as establishing that if they were there they were quivering too much to be observed. "What's the difference?" he asked with some scorn. (Some day I would like to give a talk on "My life with Fisher.") So I found a difference: the same harmonic two-dimensional solid that has no crystalline order can be shown by a simple modification of the Peierls argument to possess long range orientational order. This observation appeared in my paper. It was noticed by David Nelson, who then developed with Bert Halperin their well known theory of orientational ordering in two dimensions. Some years later I had occasion to chide the author of an article that asserted that Halperin and Nelson were the first to note that two-dimensional solids could have orientational order without positional order. I said that if he had asserted that they were the first to do something interesting with the observation I could not agree more, but since all he was talking about was who noticed it first, that had to be me, dammit. He fired right back, admitting that Halperin and Nelson weren't the first. Didn't you know, he said, it was Landau! Tough news for one struggling to qualify as a $4\frac{1}{2}$.

I have to say that in preparing this talk I have searched again for that Landau reference and cannot find it. It is probably buried somewhere in some edition of the *Statistical Physics*, but my hope has been rekindled that just maybe it isn't there after all and my credentials as a $4\frac{1}{2}$ remain secure.

I puttered around for a few more years in the theory of phase transitions, a major domain of Landau's, of course, but I didn't run directly into him very much until superfluidity was discovered in ^{3}He. It appeared to be the long-sought-for BCS pairing, but not of the isotropic *s*-wave variety found in metallic superconductors. I wrote down and had a lot of fun trying to extract information from the Landau theories of *p*-, *d*-, and *f*-wave pairing, once again playing Landau's wonderful game of exploiting nothing but the symmetries, but hitting those for all they were worth.

I then became worried about the character of the superfluidity in the so-called *A*-phase. Everybody said there was a superfluid velocity that was the gradient of a phase, but the order parameter had nodes and there was no sensible way to define a phase. Landau, I remembered, didn't like the phase anyway. How would he have handled it? I immersed myself in Landau's work on superfluidity in ^{4}He. I announced a series of lectures on Landau's derivation of two fluid hydrodynamics. And as I expounded every step of the argument, I generalized it from ^{4}He to ^{3}He-A. And at every stage it worked! The symmetries alone continued to tell you everything you needed. I began to feel an almost mystic kinship with Landau, as I followed this path he had laid out for me with almost magical vision, thirty-five years earlier.

I emerged with a funny conclusion. In ^{3}He-A the curl of the superfluid velocity was not in fact zero. But it was not unconstrained either, being related to an apparently independent degree of freedom – an axis of anisotropy – possessed by the more intricate ^{3}He-A order parameter. This relation implied a new kind of interplay between superfluid flow and the more liquid-crystal-like textural motions. I announced this conclusion at the last lecture in the series and one of my colleagues, not noted for his propensity to lavish praise, said to me afterwards – I still cherish his words – “Landau would have been proud of you.” As a $4\frac{1}{2}$ I doubt the validity of his conclusion, but the moment is unquestionably the other high point in my life with Landau.

My most recent interest has been in icosahedral ordering in the remarkable quasicrystalline alloys discovered by Schechtman in 1984. Though I have never managed to do physics for long without discovering Landau looking over my shoulder, this time I didn't expect to meet him. In the 1958 edition of the *Quantum Mechanics* in the chapter on symmetry, with that mixture of brevity and thoroughness so characteristic of Landau and Lifshitz they swiftly survey all the finite subgroups of $SO(3)$. When one comes to the icosahedral group one reads: “These groups are of no physical interest, since they do not occur in Nature as symmetry groups of molecules.” A vacation from Landau! Well, maybe yes, maybe no. In checking this out I grabbed the 1976 edition instead, and found at the same point: “These groups occur only exceptionally in Nature as symmetry groups of molecules.” And one can, of course, make Landau theories of quasicrystalline ordering, if only to explore conceptually the question of whether the ground state of positionally ordered matter has to be crystalline. And of course I did.

I cannot close a survey of my life with Landau, without remarking on my life with Landau and Lifshitz. I did once spend several days with Evgenii Lifshitz, and we hit it off at once. He denounced me for not having read Khrushchev's memoirs, which had just been published in the West, muttering in the friendliest way that I should learn to take advantage of the privileges I didn't deserve to have (and should also please stop playing ragtime on the piano and work harder on Bach). We had a wonderful time together.

Because, in the Landau manner, Landau and Lifshitz go straight to the heart of the matter with only the slightest bow to the conventional preliminary pieties, it takes a formidable concentration and self discipline to learn a subject from them. But if one has been gently led into an area by lesser guides, a subsequent study of Landau and Lifshitz can be profoundly illuminating. I remember the first time I had to teach classical mechanics struggling painfully to get Hamilton Jacobi theory out of the sprawling exposition it gets in Goldstein, before finding it efficiently packaged in one formidable page of Landau and Lifshitz's *Mechanics*.

Just before my solid state physics book with Neil Ashcroft appeared in Russian, we both happened to be at a winter school in Poland where we met Kaganov, who was editing the translation. He invited us in the course of a roaring party to write a short preface. Thoroughly besotted, Neil and I retired to a corner of the room and wrote a moving hymn of praise to Landau and Lifshitz, which we signed and handed to Kaganov without keeping a copy for ourselves. It must have come out more or less all right because a year later I received a very moving letter of thanks from Lifshitz. I thought it appropriate to quote our preface as the conclusion of this chutzpahlogical survey of my life with Landau. This proved not so easy to do, since when I attempted to translate the Russian back into English I realized that either my dictionary was not up to the task, or our original draft had been subject to further editing, or we had consumed even more vodka at the time than I remembered. In any event, the version in the book seemed slightly off key, but guided by the Russian, I think I have reconstructed what we actually wrote at the party:

> Were authors permitted to dedicate translations, we would dedicate the Russian edition of our book to E. M. Lifshitz and the late L. D. Landau. Their remarkable books have set a standard of rigor, clarity, depth, and breadth that will long remain unsurpassed. Had they written a solid state physics text there would have been no reason for us to write ours. Now that our book is available to Soviet students we hope it will begin in some small way to repay the enormous debt that we and all our students owe to Landau and Lifshitz.

I would also like to think that my small act of chutzpahlogy tonight, the homage of a $4\frac{1}{2}$ for a 2, is a similar modest (or, you may think, immodest) repayment of the incalculable debt that I, as a physicist, owe to Landau.

5

What's wrong with this Lagrangean?

A few months ago I found myself living one of my milder versions of hell, trapped on a flight to Los Angeles, having forgotten to bring along anything to read but *Physical Review Letters.* Finishing the two articles that had inspired me to stuff it in my briefcase before we even reached the Mississippi, I decided to make the best of a bad thing by taking the opportunity to expand my horizons. Scanning the table of contents, I was arrested by a title containing the word "Lagrangean."

Funny, I thought, it's not often you see misprints so blatantly displayed. But when I turned to the article, there it was again, "Lagrangean," in the title and scattered through the text. Well, I thought, an uncharacteristic failure of the copy editing process. The authors were foreign and apparently didn't know how to spell. Copy editors aren't physicists, the word is surely in few if any dictionaries, and so it slipped through.

But I had nagging doubts. Easily resolved, I thought: You can't write an article in theoretical particle physics without a Lagrangian, so I can check it right now. Well, it turns out to be not quite that easy. To be sure, you can't do particle physics without a Lagrangian, but you don't have to call it anything more than L, and many don't. Nevertheless, I found a Lagrangian, fully denominated, in one more article, and there it was, shimmering derisively before my eyes again: "Lagrangean."

Now I am not a man of great self-confidence, and my secretary will testify that I am a rotten speller. Was I fooling myself? Could "Lagrangean" be right, and my conviction that it should be

"Lagrangian" an orthographic hallucination induced by the absence of better things to read, like a mirage in the desert? Please ask yourself dear reader, before reading on: Would it have startled you?

When the plane landed in Los Angeles, I tooled up the freeway to the house of my hosts and breathlessly asked, "How do you spell 'Langrangian'?"

"I dunno," said he, but she said without hesitation, "L-A-G-R-A-N-G-I-A-N," and I felt hope for my sanity. A quick tour revealed that every book in the house on mechanics and field theory spelled it with an *i*. I *was* sane! But what was going on at *Physical Review Letters*?

The next day at UCLA and the following day at Santa Barbara, I asked almost every physicist I met how to spell "Lagrangian" (all got it right) and whether they had ever noticed *Physical Review Letters* spelling it wrong (none had). A transitory anomaly, I thought – an accident limited to the issue I happened to put in my briefcase. But when I had a free moment I went off to the library, just to make sure.

This is what I discovered: *Physical Review Letters* has been systematically misspelling "Lagrangian" with an *e* instead of an *i* since the middle of 1985. At the start of July and earlier it is "Lagrangian"; by the end of the month and thereafter it is uniformly "Lagrangean." (In the interior of July 1985 it oscillates.) They have been doing it for over two years, and nobody I asked had noticed! Nobody I have asked since has noticed! Have you noticed?

The disease is confined to *Physical Review Letters*. As far as I can tell *Physical Review* in all its multitudinous varieties is still spelling the word correctly.

I am publishing this discovery here for the first time. I claim exclusive credit for it, my extensive random samplings having led me to conclude that nobody else ever noticed "Lagrangean" during the entire two and a half years it has been lurking in the pages of *Physical Review Letters*. My discovery raises at least two serious questions, of which I save the more disturbing for last.

Question 1. What is going on here? Why is *Physical Review Letters* misspelling "Lagrangian"? One can invent theories. To be sure, the man's name was Lagrange, ending, as any undergraduate can tell you, with an *e*. But if you write "Lagrangean," then shouldn't you pronounce it "luh-GRAN-jin," and doesn't everybody actually say "luh-GRAN-jee-in"? Doesn't "Lagrangean" lead unavoidably to "Hamiltonan," which gives me, for one, a case of the giggles, and certainly has never been sighted in the pages of *Physical Review Letters* or any other journal of repute? Ah, but "Hamilton" doesn't end with *e*. Well, what about people who do end with *e*? Try adjectivizing them. Don't you want to turn their *e* into an *i* before the *an*? Or do you?

Such talk, fascinating as it can become, utterly misses the point. English spelling is entirely irrational. Theorizing about it is a form of what Einstein called "brainschmaltz." There are no rules, only precedent. And precedent demonstrates unanimously, overwhelmingly and unambiguously that for at least a quarter of a century the English word has been and remains "Lagrangian."

I devoutly hope the answer is that a bug crept into their spelling checker in the summer of 1985, but I fear the worst, and I therefore here declare that *Physical Review Letters* has no right to tamper with established usage. One can only hope the editors will soon come to their senses.

Question 2. The more disturbing question: Why has nobody noticed? Why did this aberration lie undiscovered for more than two years, only coming to light because one careless man allowed himself to fall into a dreadful trap that any prudent person would have taken simple measures to avoid? Can it be that physicists no longer know how to spell? No, because when I asked a random sample, they all spelled it "Lagrangian." Can it be that they are all speed readers, zooming on to the next word as soon as they get past the opening "Lag"? I don't think so. It seems to me that very fast readers take in whole lines at a time, and when you look at a whole line and you know how to spell, what you see glaring out at you, defiantly thumbing its nose, is "Lagrangean." We do know how to spell. We do see what we read. I can think of only one other explanation, but it is an explanation so alarming, so staggering in its implications, that I hesitate to give voice to it:

Can it be that nobody any longer reads *Physical Review Letters*?

We've known for some time that, roughly speaking, nobody any longer reads anything but preprints, the archival journal of choice, which for many years now has been *Physical Review Letters*, and secondary references cited in these two primary sources. But the preprints have been coming thicker and faster. And *Physical Review Letters* now publishes almost as many pages each month as *all* of *Physical Review* did back in 1956, when I was starting graduate school. (And at that time *Physical Review* included the letters as well as containing within one set of covers all of *A*, *B1*, *B15 I*, *B15 II*, *C*, *D1* and *D15*). Yet slim as it was, and few as the other journals were, back in those easygoing days *Physical Review* was widely known as "the green plague." *Physical Review Letters* is now as big as the green plague of the 1950s, and the white plague (preprints) is even bigger. Is it then indeed possible that people have stopped reading it?

Few, of course, when asked about their reading habits will give you a straight answer, but I submit for public discussion what has to be regarded as a very powerful piece of evidence that the pages of *Physical Review Letters* are now examined no more than any of the other hundred thousand or so pages that pile up each month in our physical science libraries. I submit that whoever decided to start systematically misspelling "Lagrangian" was unwittingly (or could it have been wittingly?) conducting a beautiful experiment that could not have been more ingeniously contrived to get an honest measure of how carefully people actually look at *Physical Review Letters.*

The results of that experiment are disconcerting, with implications for at least two major problems that we have not adequately faced as a profession: the disaster looming over science libraries, and therefore over science itself, as a result of the irresponsible way we have allowed scientific journals to proliferate; and, not unrelatedly, the lamentable decline in the quality of scientific writing.

These problems are addressed in the three essays that follow.

Postscript

Just before this article appeared, while browsing through the *Encyclopaedia Britannica*, I discovered that Lagrange was originally named Lagrangia. It was too late to get this into the text, but I managed to append it to the author identification blurb: "David Mermin is a professor of Physics at Cornell University. . . . He has recently learned that Joseph-Louis Lagrange was born in Torino and named Giuseppe Luigi Lagrangia." I subsequently received an angry letter from Turin denouncing "Lagrangia" as an abominable piece of fascist revisionism and indeed, going back to the *Britannica* I found that a biography from the early 1940s was cited as a source. My inquiry to the *Britannica* was acknowledged with thanks and a promise of details to follow (which has yet to be fulfilled). At least one person from Turin I later met was rather sure that "Lagrangia" was authentic. All this is of course, entirely beside the point, as is the fact (widely unnoticed) that *Physical Review Letters* once again spells "Lagrangian" the same way as everybody else.

6

What's wrong with this library?

> An extrapolation of its present rate of growth reveals that in the not too distant future *Physical Review* will fill bookshelves at a speed exceeding that of light. This is not forbidden by relativity, since no information is being conveyed.

I first heard this joke from Rudolph Peierls in 1961. Since then the size of *Physical Review* has doubled, doubled again, and is on the verge of completing its third doubling. *Physical Review Letters* is nearly as big as *Physical Review* was in 1961. The journals of other physical societies have undergone comparable expansions, as have the numbers and sizes of the commercially published physics journals.

I was recently asked to address a group of Cornell alumni and librarians on how the library is used by a typical physical scientist. I did point out that they had selected a distinctly oddball specimen, but they persisted. So to give them a sense of the calamitous conditions we have created, I went systematically through the current periodicals section of the Cornell Physical Sciences Library to count how many journals I felt I ought to look at but didn't. My criterion was stringent. A journal made my list only if it would be downright embarrassing to admit that I never looked. I had to be able to imagine a colleague responding to my confession with amazement: "You never look at *X*?!" Counting only once those with multiple versions (*A*, *B*, *C*,...) I still found 32. In one case, uncertain whether an admission that I never looked would occasion a blush or merely a nervous giggle, I randomly opened a random issue and found an article I had not seen before, expanding interestingly upon earlier work of my own. *That* information had not been conveyed.

It is in this context that we must view the resistance of many scientists to current efforts by library administrators to make draconian cuts in their subscription lists. Keeping up with everything is driving the libraries into bankruptcy. Early signs of this problem and how physicists might respond surfaced over ten years ago. My colleague Ken Wilson and I, physics department representatives on the library committee and therefore conscious of the looming catastrophe, responded to announcements of two new physics journals by writing a letter to *Physics Today* (March 1976, page 11) announcing that our library could not afford any more subscriptions. We warned potential authors that if they published in the two newcomers they would therefore be unread at Cornell, and urged for the sake of science libraries everywhere that people should stop writing for, lending their names to the editorial boards of and instructing their libraries to subscribe to these and other unnecessary new publications.

From around the world we were attacked with a fury I couldn't have imagined. Many held us to be the running dogs of Jim Krumhansl, also at Cornell and at that time editor of *Physical Review Letters*, in a blatant conspiracy to stifle all competition. Nobody believed for a minute that our concern was only, as we stated emphatically, for the survival of our library. As it turned out, one of the journals died within a year or two, for reasons unrelated to our kamikaze attack; the other still exists and, I'm ashamed to admit, the last time I checked Cornell was receiving it.

Now, 12 years later, things have worsened. Our physical sciences library has been given a budget we can live within only by cutting our already pruned subscriptions by another third, and cuts on this scale are not uncommon at other libraries. Scientists have two common responses to this threat to the integrity of their libraries:

> A university is obliged to maintain its science libraries at the levels its scientists require because the library is included among the indirect costs of sponsored research for which the granting agencies reimburse the university.

This argument plunges its unwary proponent into a snake pit of accounting tricks and countertricks, ringed about with ethical analysis that makes Kant read like Ann Landers. In its most coherent form the opening salvo goes like this: Journals are as important a part of doing research as magnets or liquid nitrogen. Rather than do without, we would include their cost in the budgets of our grants, but we cannot because library costs have already been taken into account in the indirect-cost calculation. The university has already been furnished the means to provide for our library needs. If those funds are insufficient the indirect-cost formula should be renegotiated. If they *are* sufficient... (here the analysis flows into various vituperative channels that need concern us no further). This argument is rapidly being rendered obsolete, at least in my own field of condensed matter physics, by the increasing inability of the agencies to provide support even at the level of magnets and liquid nitrogen, but this seems not yet to have caught up with proponents of the Argument from Overhead.

> Alright [damn you], cut. But be sure to dovetail your cutting with other libraries in the city [state, region] so any journal can still be found within 25 [50, 200] miles from here and rushed over within 4[8, 24] hours.

This short-term fix will reduce the circulation of many journals, and we know how publishers with a captive audience will respond (and never were audiences more captive than university libraries to science journals): by raising their prices. Under this scheme each library will soon pay as much as before for its diminished collection.

Both the Argument from Overhead and the Dovetail Fix miss one central fact: There are too many journals. The problem is not how to persuade library administrations or granting agencies to produce funds to keep them all at our fingertips, nor is it how to keep them readily available from a wider library pool. The problem is how to get rid of them.

It is a little hard to figure out how professionally rational people managed to land themselves in such a mess. Part of the blame must be assigned to the practice of good science libraries, now finally abandoned, of subscribing to everything regardless of cost, thereby offering publishers an irresistible incentive to launch new journals regardless of need. (One of the less attractive manifestations of high-temperature superconductivity is the demonstration it affords that the era of such launchings has not entirely died.) But why did we collaborate in this milking of our profession by sending publishers our articles and agreeing to be on their editorial boards whenever they saw another opportunity to rip us off? Vita enhancement is surely an insufficient motivation. In some cases it seems to have been nothing more than shortsighted penny-pinching. I have been told, for example, that it is impossible to do research in theoretical particle physics without a commercially published journal for whose two dozen neat little volumes we currently pay $4000 a year. Why do so many particle theorists publish there rather than in *Physical Review D*? You guessed it: no page charges.

We are probably stuck with that pricey little item for good, but most of the redundant journals have yet to reach the level of an addiction. Getting rid of them requires precisely the opposite of the dovetailing strategy. When my library decides to cancel a journal, far from it being important for neighboring libraries to retain it, they and most other libraries must cancel too. The journal will then expire because the few remaining subscribers, abused as they have been in the past, will not put up with a further massive increase in price, and because nobody wants to publish in a journal that significant numbers of their colleagues will not even have the opportunity to feel embarrassed at not examining.

How can we bring this about? To some extent it will happen automatically. As physicists across the country are asked by their libraries to specify what is to be dropped, provided we are not seduced by the dovetailing fallacy, the various lists of journals we can survive without will have many members in common, if there is any objectivity in our science. Harder to deal with will be those many journals with distinguished boards of editors in which 5 percent of the contents are respectable and even interesting. How are their numbers to be reduced?

Various schemes come to mind that I am not worldly enough to assess the feasibility of, much less bring to pass:

A special evening session at every major APS meeting at which representatives of the library committees at universities and government and industrial laboratories would get together and agree on a common list of recommendations for the axe in that subfield.

If the powers of the APS considered such a spectacle unedifying, what about providing a central clearing-house of information on recent cancellations? Libraries would inform the clearinghouse of their cancellations, and a periodic newsletter (*Look who's not subscribing to what!*) would be sent to libraries and other interested parties, listing recent cancellations by journal and by library. To achieve the desired end it would probably suffice to circulate widely the cancellations of 30 or so major university libraries and a dozen or so industrial and government ones.

If the APS will not participate even in this effort (and maybe it shouldn't, since, after all, it publishes journals too) then departmental library committees could do it informally by swapping lists of recent cancellations. "Thought you might like to know that last month we decided we could do without the following 37 and canceled. What have you been up to lately?"

Restraint on the part of individuals is also essential. The default response to a request to be on the editorial board of a new publication should be an emphatic no, accompanied by a reasoned statement of why the journal should not be started at all. You should review all journals on whose editorial boards you currently sit, asking whether the world would be significantly worse off if the papers appearing there were published elsewhere. In most cases you will have to admit that the world would on balance be better off. You must then resign immediately, explain why and send your distinguished fellow editors copies of your resignation for their moral improvement.

Most importantly, however, and perhaps most painfully, we should all think twice before writing yet another article. For although significant economies can certainly be achieved by channeling the great flood of papers into fewer vessels, we (and the agencies that give us grants and the authorities that approve our promotions) have forgotten that the aim of publishing articles is to communicate. The current crisis of our libraries offers us a rare opportunity to draw back from our relentless march toward that dreaded point beyond which no information is conveyed.

7

What's wrong with this prose?

I write bleary eyed and disheartened, after a long proofreading session mainly devoted to inserting into the galleys calls for the restoration of what was capriciously and destructively altered in the editorial offices of *Physical Review*. I proofread simply by reading the galleys, without reference to the original manuscript. My writing is a process that does not converge; I cannot read a page of my own prose without wanting to improve it. Therefore when I read proofs I entirely ignore the manuscript except to check purely technical points. Proofreading offers one more shot at elusive perfection. Proceeding in this way, I come to the end of a paragraph with a lurching sensation. The last sentence seems to be a *non sequitur*. Can I be failing to get my own point? Turning to the copy edited manuscript, I find a marginal message: "Author: Please note that we discourage single-sentence paragraphs." As an application of this principle, one short emphatic paragraph has been attached to the end of another, to which it is entirely unrelated. If you set asunder what *Physical Review* has joined, it makes sense again.

What is the justification for such a rule? Excessive use of single-sentence paragraphs blurs the distinction between the sentence and the paragraph, makes for a visually unattractive page, and becomes boring. But the occasional single sentence paragraph is a valuable device. It gives a pause in the rush of thought, it focuses attention, and it can contribute powerfully to the rhythm of the prose. The Constitution of the United States of America, whose prose Warren Burger enjoined us to admire in its 200th anniversary year, is chock full of beautiful single-sentence paragraphs. A blanket prohibition is absurd, and enforcing it by paragraph grafting is almost certain to do violence to the clarity and even the meaning of a well written essay. So I go through the galleys restoring the three or four indigestibly merged paragraphs, adding my own marginal messages ("Editor: We discourage gratuitous confusion") in the hope that my counter-instructions will not be ignored.

A bit later I come to a reference to nature. "Nature herself," I remember writing, "has proved to be quite unambiguous...." The galley reads "Nature has proven quite unambiguous..." Not bad, I think, getting rid of that unnecessary "to be" – should have spotted it myself. But then I notice that nature has been depersonified. Why can't nature be "she"? Could "herself" have been sacrificed in an enlightened attempt to exorcise unconscious sexism from the pages of *Physical Review*. No. [*Author: Please note that we discourage one word sentences.*] The desexing of Mother Nature is explained by "Author: Please note that the editor feels this wording to be more literal, and therefore preferable." The note refers me to other applications of the same rule to my manuscript: the adjective has been deleted from a reference to a "charming monograph" and "aficionados of ring theory", has become "ring theorists". The first alteration has deprived the reader of the information that the work in question is uncharacteristically readable for a monograph on number theory; the second eliminates the information that ring theory is not part of the everyday mathematical equipment of most physicists, and also introduces an absurdly inappropriate pomposity (compare "evolution theorists" or relativity theorists" or "group theorists").

The next thing I run into is "Author, please place only a word or short phrase rather than a whole sentence in italics." Well, OK. I can see that whole sentences in italics might make for a blotchy kind of page, particularly if there are lots of equations around. But occasionally it can be quite useful to call attention to a central point by putting it in italics. I maintain that anything you can do to help the reader follow your argument is worth doing. Nevertheless, I'm willing to forego excessive use of the italic option for the sake of a neater page. But what have they done at *Physical Review*? They haven't removed *all* the italics; selected (God knows how) words in those formerly italicized sentences have been left in italics, with almost uniformly preposterous results. (My proofs sport about ten such *sentences*, all *reading* like this one; I freely *admit* that I probably got *carried* away with italicized sentences, but surely the cure is worse than the *disease*).

Stranger still, in the caption of a geometric figure the assertion that the straight line joining point B to point F has the same length as the straight line joining point A to point F, which appears in the manuscript as "$BF = AF$" has been transformed into "$d(BF) = d(AF)$ (where d is the distance)". This violates three cardinal rules at once: (1) Do not introduce unnecessary notational complexity; (2) Do not introduce unnecessarily unconventional notation; (3) Do not make lengthy that which is brief.

And so it goes. *Physical Review* is certainly not the only practitioner of destructive copy editing. *Scientific American* is notorious for elephant walking over the writing that enters its offices, systematically pounding it into homogeneous soporific mush. Even *Physics Today* which publishes some of the better prose in the scientific literature, is not without its foibles.[1] Were you, for example, reviewing a concert for *Physics Today* you would be required to talk about "Wolfgang Mozart's *Jupiter Symphony*" because, I can only imagine, the reader might be under the impression that Leopold wrote one too. It would have to be Johann Goethe's *Faust* and Sandro Botticelli's *Primavera*. Everybody you mention in *Physics Today* has to have a first name. This is absurd and can also be destructive of good writing, introducing the literary equivalent of a hiccup into a smooth sentence, or raising in the reader's mind such spurious questions as "Why *Werner* Heisenberg; was there another I didn't know about?"

[1] I am told they are thinking of reforming.

Why am I telling you all this? Surely you all have stories of assaults on your manuscripts as irritating as mine. Precisely. I raise the matter to urge you to fight back. This savaging of our prose – this obliteration of our human individuality – has something important to do with one of the great failures of science in our time: the virtual disappearance of just plain readable – never mind humane – scientific prose. This is a calamity for science, and not only because it makes the practice of science much less fun. Bad thinking is vastly easier to cover up if you're allowed to get way with and even encouraged to produce bad writing.

Among the principles underlying these examples of copy editing is the intention to eliminate any trace of a human author. The inevitable result is a bland uniformity. By making the point that anything remotely lively, idiosyncratic, or quirky will be eliminated, *Physical Review* deprives an author of any incentive to write interestingly, and worse, makes it very much more difficult for an author to provide gracefully the kinds of emphases and signposts without which scientific exposition can become virtually unintelligible.

Eliminating the artificial obstacles to decent scientific prose erected by *Physical Review* will not in itself ensure the return of clean and vibrant writing to its pages, but as long as the copy editing process continues to emasculate or defeminize our texts, there is no hope that we can breathe the life back into scientific writing or persuade our students that writing well is a worthy and even noble endeavor. The final result of our efforts as scientists, is, after all, not a table of data, a set of equations, or the output of a computer. It is an essay, a piece of expository prose. That's what grant officers, promotion committees, and biographers care about and for once they're right.

So fight back. Restore the humanity to your bowdlerized text when the galleys arrive. Victory does not come easily, but it will never come to those who refuse to fight. I changed "monograph" back to "charming monograph" in the proofs, and I write now, over a year since I began this essay, to report the results. I got a call from *Physical Review*.

"About that monograph...," the man said.

"Yes?"

"How would you like 'interesting monograph' or 'important monograph'?"

"Well," I said, "as a matter of fact it isn't *terribly* interesting. And *nobody* could honestly say that it was important." Long pause. "But it *is* charming."

"Oh," he said, "I see."

And it stayed "charming."[2] Therefore do not hesitate to write interesting, readable, lively, intelligible articles. It is your duty to do so. And when the proofs come back duller, clumsier, and more ambiguous than the manuscript you sent in, restore the life to those galleys, and be calm but firm when the phone rings. You will not only have more fun that way, but you will also be contributing to the good fight to reverse the sad and dangerous decline of scientific discourse in our time.

[2] *Phys. Rev.* **B35**, 5495 (1987).

8

What's wrong with these equations?

A major impediment to writing physics gracefully comes from the need to embed in the prose many large pieces of raw mathematics. Nothing in freshman composition courses prepares us for the literary problems raised by the use of displayed equations. Our knowledge is acquired implicitly by reading textbooks and articles most of whose authors have also given the problem no thought. When I was a graduate teaching assistant in a physics course for non-scientists, I was struck by the exceptional clumsiness with which extremely literate students, who lacked even the exposure to such dubious examples, treated mathematics in their term papers. The equations stood out like dog turds upon a well manicured lawn. They were invariably introduced by the word "equation" as in "Pondering the problem of motion, Newton came to the realization that the key lay in the equation

$$F = ma.\text{"} \qquad (1)$$

To these innocents equations were objects, gingerly to be pointed at or poked, not inseparably integrated into the surrounding prose.

Clearly people are not born knowing how to write mathematics. The implicit tradition that has taught us what we do know contains both good strands and bad. One of my defects of character being a preference for form over substance, I have worried about this over the years, collecting principles that ought to govern the marriage of equations to readable prose. I present a few of them here, emphasizing that the list makes no claim to be complete. We are constantly assaulted by so many egregious violations of even these simple precepts that I offer them in the hope that a few sinners – not only writers, but copy editors, publishers of journals, and even the authors of the mathematics subsections of literary style manuals – may read them and repent the error of their ways, or even be inspired to further beneficial studies of the sadly neglected field of mathematico-grammatics.

Rule 1 (Fisher's Rule). This rule, named after the savant who reprimanded me for abusing it when I was young and foolish, simply enjoins one to *Number All Displayed Equations.* The most common violation of Fisher's rule is the misguided practice of numbering only those displayed equations to which the text subsequently refers back. I call this heresy Occam's Rule. Back in the days of pens, pencils, and typewriters, use of Occam's Rule was kept under control by the pain of having to renumber everything by hand whenever it was deemed wise to add a reference to a hitherto unremarked upon equation. One often encountered papers displaying the results of the ungainly Fisherian–Occamite compromise: Number all displayed equations that you think you *might* want to refer to. Now that automatic equation numbering macros can act upon symbolic names, the barrier to full Occamism has been removed, and it is necessary to state emphatically that Fisher's Rule is for the benefit not of the author, but the reader.

For although you, dear author, may have no need to refer in your text to the equations you therefore left unnumbered, it is presumptuous to assume the same disposition in your readers. And though you may well have acquired the solipsistic habit of writing under the assumption that you will have no readers at all, you are wrong: there is always the referee. The referee may well desire to make reference to equations that you did not. Beyond that, should fortune smile upon you and others actually have occasion to mention your analysis in papers of their own, they will not think the better of you for forcing them into such locutions as "the second equation after (3.21)" or "the third unnumbered equation from the top in the left hand column on p. 2485." Even should you solipsistically choose to publish in a journal both unrefereed and unread, you might subsequently desire (just for the record) to publish an erratum, the graceful flow of which could only be ensured if you had adhered to Fisher's Rule in your original manuscript.

Rule 2 (Good Samaritan Rule). A good Samaritan is compassionate and helpful to one in distress, and there is nothing more distressing than having to hunt your way back in a manuscript in search of Eq. (2.47) not because your subsequent progress requires you to inspect it in detail, but merely to find out what it is *about* so you may know the principles that go into the construction of Eq. (7.38). The Good Samaritan rules says: *When Referring to an Equation Identify It by a Phrase as Well as a Number.* No compassionate and helpful person would herald the arrival of Eq. (7.38) by saying "inserting (2.47) and (3.51) into (5.13)..." when it is possible to say "inserting the form (2.47) of the electric field **E** and the Lindhard form (3.51) of the dielectric function ε into the constitutive equation (5.13)...", To be sure it's longer this way. Consistent use of the Good Samaritan Rule might well increase the length of your paper by a few percent. But admit it: your paper is probably already too long by at least 30% because you were in such a rush to get it out you didn't really take enough care putting it all together. So prune elsewhere, but don't force your poor readers – you really *must* assume you will have some, or it is madness to go on writing – to go leafing back when a few words from you would save them the trouble.

Admittedly sometimes an equation is buried so deep in the guts of an argument, so contingent on context, so ungainly in form that no brief phrase can convey to a reader even a glimmer of what it is about, and anybody wanting to know why it was invoked a dozen pages further on cannot do better than to wander back along the trail and gaze at the equation itself all glowering and menacing in its lair. Even here, the mere attempt to apply the Good Samaritan Rule can have its benefits. If the nature of the equation is inherently uncharacterizable in a compact phrase, is the cross reference really necessary? Indeed, is the equation itself essential? Or is it the kind of nasty and fundamentally uninteresting intermediate step that readers would either skip over or, if seriously interested, work out for themselves, in neither case needing to have it appear in your text? If so, drop it. You will then have to revise the argument that referred back to it, but the chances are good that clarity will benefit from not having at the heart of the argument an uncharacterizable monster of an equation.

Rule 3 (Math is Prose Rule). The Math is Prose Rule simply says: *End A Displayed Equation with a Punctuation Mark*. It is implicit in this statement that the absence of a punctuation mark is itself a degenerate form of punctuation, which, like periods, commas, or semicolons, can be used *provided it makes sense.* For unlike the figures and tables in your article, unlike dog turds on a lawn, the equations you display are embedded in your prose, and constitute an inseparable part of it. The detailed theory of how equations are to be viewed as prose need not concern us here. Sometimes they function as subordinate clauses, the equals sign being the verb; sometimes they appear as substantive phrases, like a list of the contents of a room; sometimes, regrettably, they must merely be presented to the reader as objects like quotations (but with the convention that quotation marks are not required [except in the rare case that Math is Prose requires it, as in Eq. (1) above (which I never dreamed I would be referring back to when I first put it into this essay)]).

Regardless, however, of the often subtle question of how to parse the equation internally, certain things are clear to anyone who understands the equation and the prose in which it is embedded. Thus the end of the equation may or may not coincide with the end of the sentence in which it stands. If it does, then the equation should end with a period or, rarely, if the equation terminates an interrogative sentence, it should end with a question mark. (Having now succeeded in publishing an equation requiring a quotation mark, it remains my dream to publish an article with an equation that requires a question mark; somehow I haven't got around to it.) If the equation terminates a clause, or is part of an extended list, then it should end with a comma or semicolon. Only infrequently is no punctuation required, as, for example, in "Only when

$$\sum_{i=1}^{N} f(x_i) = 0 \tag{2}$$

is it impermissible to divide by this sum."

We punctuate equations because they are a form of prose (they can, after all, be read aloud as a sequence of words) and are therefore subject to the same rules as any other prose. To decree that every sentence should end in a period *unless* the sentence terminates in a displayed equation is grotesque. (If you disagree, try the rule that every opening quotation mark must be followed by a closing one unless the quotation terminates in an equation.) But one does not punctuate equations only because it is ugly not to; more importantly, punctuation makes them easier to read and often clarifies the discussion in which they occur. Acquiring the habit of viewing an equation not as a grammatically irrelevant blob, but as a part of the text fully deserving of punctuation, can only improve the fluency and grace of one's expository mathematical prose.

Most journals punctuate their equations, even if the author of the manuscript did not, but a sorry few don't, removing all vestiges of the punctuation carefully supplied by the author, thereby unavoidably weakening the coupling between the math and the prose, and often introducing ambiguity and confusion. I'm sorry to say that *Physics Today* is guilty of this practice. To be sure, their use of equations is sufficiently light that this does not inflict substantial hardship on their readers, but it greatly undermines the role they so commendably play in other respects as models of good writing about hard science. May the appearance of Eq. (1) above signal the start of a new and better tradition.

We should strive, more generally, to make errant journals mend their ways. It is easier than you might think. One of my students and I once did a piece of work that required us to lead the reader (or at least, we know for a fact, the referee) through unavoidably dense thickets of equations. Unfortunately the otherwise obvious journal for our paper systematically violated the Math is Prose Rule, so in our letter of submission we emphasized that the punctuation in our equations was essential for the comprehensibility of our argument. The letter of acceptance, however, informed us that the publisher adhered in this and all its other journals, as well as in its books, to a firm policy of never punctuating equations. In that case, we wrote back, just return the manuscript and we'll send it somewhere else. After a long pause we were informed that at a meeting of the board of directors of the publishing firm a special dispensation had been granted to our paper, and indeed, it appeared with punctuated equations.[1]

Fortunately Fisher's Rule and the Good Samaritan Rule don't require assent from boards of directors, so you have nobody to blame but yourself if your papers don't observe them; you can mend your ways right now. At a minimum you will make life much easier for an overworked referee, and with luck you might even have a few happily undistressed readers.

[1] *Found Phys.* **14**, 1 (1984).

9

What's wrong with these prizes?

But "glory" doesn't mean "a nice knock down argument."
Alice to Humpty Dumpty.

It seems to me evident that the system of prizes, honors and awards in physics has run completely amok, absorbing far too much of the time and energy of the community in proportion to the benefits conferred. Yet nobody complains. Every month *Physics Today* routinely announces the latest crop of winners, all the major American Physical Society meetings have sessions to bestow prizes, the APS directory continues to distinguish the asterisked from the unasterisked, and nobody ever complains. Why?

To ask the question is to answer it. Indeed, merely by publishing the above paragraph I have probably already irreparably blemished my reputation in the profession, and if *Physics Today* has actually printed this column I imagine it can only have been after heated and prolonged editorial debate. Much of this essay, in fact, sat aging in my computer in a directory with highly restricted access for almost two years. It was finally sprung loose by the 1988 Presidential campaign, which filled me with so intense a loathing for those who hesitate to speak provocative truths that I can no longer restrain myself. Here I go.

Why does nobody ever complain? Nobody complains because there are two categories of physicists: those who have won prizes and those who have not. Winners cannot criticize the system. It would be rude to the donors of their prizes. It would be offensive to the committee that selected them and the people who wrote letters on their behalf. It would be a vulgar display of bad taste. It would be unseemly to criticize a system one has benefited from before others have had their chance to win.

But neither can nonwinners criticize the system. It is not that a public attack on, for example, the absurdity of election to the National Academy of Sciences might jeopardize one's own chances for immortality, for this would be a noble sacrifice. What freezes dissent for the nonwinner is that it would be perceived as sour grapes – an unbecoming outburst of petty jealousy. The only respectable stance for the nonwinner is warmly to congratulate each new crop of winners, a kind and gentle response to be sure, but one that implicitly endorses the system itself, preposterous as it is.

At this point you may well be distracted from my original contention by the question of which camp you are being addressed from. I have wrestled at some length with whether to declare myself at the outset or force the curious into a possibly quite lengthy perusal of various arcane archives. The only satisfactory solution I have come up with is to invite anyone wanting to know to send me a stamped, addressed envelope, which I promise to return with an up-to-date CV.

Interestingly enough, by leaving unspecified my own level of glorification it seems to me that I am, at least with those readers who deem it as likely that I am glorified as not, doing considerably less damage to my reputation for courtesy, tact and simple decency than I would have done had I declared myself explicitly to be either of the two (exhaustive and mutually exclusive) types. This is as close to a demonstration of quantum interference on the sociological level as I have ever encountered. But I digress.

I realized that the honor system had become a destructive force shortly after having assumed certain administrative responsibilities. Before that I had never thought much about it one way or the other, occasionally submitting essays on behalf of deserving people I thought had been overlooked, noting with pleasure the good awards, and with irritation or amusement, the bad ones. Only recently did I learn ("How innocent can you get!" you will say, dear reader – you who have known the dark side of awards longer than I, and yet have never spoken up publicly against the whole business) that these things are systematically sought after by organized campaigns, routinely consuming oceans of time and effort.

If we don't put up all of our guys, they'll win with theirs, seems to be the guiding principle. No point in disinterestedly recommending the most deserving, irrespective of institutional affiliation, for such people are already being backed by their own teams. Conversely, if we don't push our own, nobody else will. The folklore in my corner of physics is that it's the industrial laboratories that put up the most massive and systematic campaigns, but in my experience the universities have been quick to acquire the bad habits of all whom they deal with, and I wouldn't want to say who are the worst offenders.

Once you start down this path the process acquires a crazy momentum. If you have put across a winner you can't sit back and enjoy the satisfaction of a job well done. Can one rest after X gets prize A? Certainly not; 65% of all winners of the A Prize go on to receive the B Medal, half of the B Medalists become fellows of the D E of F, and it would be an irresponsible administrator who didn't go for the whole pile. Worse, as even the slightest aura of glory becomes attached to routine professional activities – for example, giving a talk at a meeting – the point of selecting people for such jobs flips from finding the best to supporting the team (which in the case of my team [but not yours] amounts to exactly the same thing).

This stampede after glory, foreign and domestic, would be a piece of harmless silliness, did it not involve such a substantial expenditure of human energy. Most of us are asked to make other judgments that, unlike the cosmetic decisions in the glory game, are of vital importance to the professional survival of our colleagues. We are asked to review grant proposals, we are asked to referee manuscripts, we are asked to evaluate colleagues for appointments to new positions or promotions. No responsible member of the profession could refuse to do these things, but most of us do so many of them that we don't do a very good job. We simply haven't the time.

I maintain that with all these serious demands on our attention, this childish scramble after glory is a frivolity we can no longer afford. How to relieve ourselves of it is less clear. It would be too much to hope for the abolition of all prizes and self-perpetuating honorary societies. The child in each of us cannot, and probably should not, be entirely obliterated. Baseball understands these things, and does them much better than we do, conferring the Most Valuable Player Award by decision of the sportswriters, leaving the players themselves to get on with more serious business. A moment's reflection on the spectacle of even the top science writers voting to select, say, the Physics Rookie of the Year reveals that this won't work. As a spectator sport, physics is a complete bust. The rules are too complicated, and the science writers can't really judge performance.

My guess is that it is up to the people who make these distinctions to save the rest of us from this frenzy of unproductive effort. It would be unfair to ask selection committees to refuse all external nominations and do the entire job themselves, though whatever else one might think about the MacArthur Awards, they do have the not inconsiderable virtue of wasting the time of relatively few in the selection process. But could not the bestowers of prizes limit to one the number of people they were willing to hear from in support of any given nomination? Suppose it were specified that there would be a preliminary screening of all letters nominating a candidate to determine which single one was to be retained, the others being destroyed without keeping any record of them or their authors. Presumably any qualified observer can summarize the nominee's accomplishments. Stirring up the mighty of the Earth to bombard committees with letters of their own repeating these data is a ritual it is high time to set aside. Let the committee make a few phone calls if it wants confirmation of the one letter.

Better yet, why can't people nominate themselves? Indeed, why not insist on it? Who, after all, is better qualified to prepare the case, and more likely to do it with verve and enthusiasm? We are already, with only a few unfortunate exceptions, the only ones who nominate ourselves for research grants or our own prose for inclusion in prestigious journals. Should I die wealthy I will endow an APS prize (probably for Theoretical Contributions to Statistical and Low-Temperature Physics by One with a Fine Prose Style). The Mermin Prize will be available only to applicants who submit an essay of no more than 500 words demonstrating explicitly and implicitly why they qualify, a list of no more than eight relevant papers, and the names of two people the committee might or might not want to consult in a phone conversation of no more than three minutes' duration. The names of all applicants will be published in the APS *Bulletin* to discourage the frivolous and install a proper humility in the serious, for the point of the Mermin Prize will be not glory but money – \$750 000 sounds good. Applications will remain valid for four years, no updating permitted, after which unsuccessful applicants will become ineligible. People with no interest in the process can go on peacefully doing physics.

I offer these views in the hope that having thus shot myself in the foot, I may encourage others to voice their opinions on what, if anything (hold on for a breathtaking swerve of metaphor), this particular emperor is or ought to be wearing. Can't we discuss this business out in the open? Or is it too much like explaining on prime-time television that it's *wrong* – never mind unconstitutional – to force people to pledge allegiance to a flag?

II.

The quantum theory

10

Quantum mysteries for anyone

We often discussed his notions on objective reality. I recall that during one walk Einstein suddenly stopped, turned to me and asked whether I really believed that the moon exists only when I look at it.

A. Pais[1]

As O. Stern said recently, one should no more rack one's brain about the problem of whether something one cannot know anything about exists all the same, than about the ancient question of how many angels are able to sit on the point of a needle. But it seems to be that Einstein's questions are ultimately always of this kind.

W. Pauli[2]

Pauli and Einstein were both wrong. The questions with which Einstein attacked the quantum theory do have answers; but they are not the answers Einstein expected them to have. We now know that the moon is demonstrably not there when nobody looks.

The impact of this discovery on philosophy may have been blunted by the way in which it is conventionally stated, which leaves it fully accessible only to those with a working knowledge of quantum mechanics. I hope to remove that barrier by

[1] *Reviews of Modern Physics*, **51**, 863 (1979): 907.

[2] From a 1954 letter to M. Born, in *The Born–Einstein Letters* (New York: Walker, 1971), p. 223.

describing this remarkable aspect of nature in a way that presupposes no background whatever in the quantum theory or, for that matter, in classical physics either. I shall describe a piece of machinery that presents without any distortion one of the most strikingly peculiar features of the atomic world. No formal training in physics or mathematics is needed to grasp and ponder the extraordinary behavior of the device; it is only necessary to follow a simple counting argument on the level of a newspaper braintwister.

Being a physicist, and not a philosopher, I aim only to bring home some strange and simple facts which might raise issues philosophers would be interested in addressing. I shall try, perhaps without notable success, to avoid raising and addressing such issues myself. What I describe should be regarded as something between a parable and a lecture demonstration. It is less than a lecture demonstration for technical reasons: even if this were a lecture, I lack the time, money, and particular expertise to build the machinery I shall describe. It is more than a parable because the device could in fact be built with an effort almost certainly less than, say, the Manhattan project, and because the conundrum posed by the behavior of the device is no mere analogy, but the atomic world itself, acting at its most perverse.

There are some black boxes within the device whose contents can be described only in highly technical terms. This is of no importance. The wonder of the device lies in what it does, not in how it is put together to do it. One need not understand silicon chips to learn from playing with a pocket calculator that a machine can do arithmetic with superhuman speed and precision; one need not understand electronics or electrodynamics to grasp that a small box can imitate human speech or an orchestra. At the end of the essay I shall give a brief technical description of what is in the black boxes. That description can be skipped. It is there to serve as an existence proof only because you cannot buy the device at the drugstore. It is no more essential to appreciating the conundrum of the device than a circuit diagram is to using a calculator or a radio.

The device has three unconnected parts. The question of connectedness lies near the heart of the conundrum, but I shall set it aside in favor of a few simple practical assertions. There are neither mechanical connections (pipes, rods, strings, wires) nor electromagnetic connections (radio, radar, telephone or light signals) nor any other relevant connections. Irrelevant connections may be hard to avoid. All three parts might, for example, sit atop a single table. There is nothing in the design of the parts, however, that takes advantage of such connections to signal from one to another, for example, by inducing and detecting vibrations in the table top.

By insisting so on the absence of connections I am inevitably suggesting that the wonders to be revealed can be fully appreciated only by experts on connections or their lack. This is not the right attitude to take. Were we together and had I the device at hand, you could pick up the parts, open them up, and poke around as much as you liked. You would find no connections. Neither would an expert on hidden bugs, the Amazing Randi, or any physicists you called in as consultants. The real worry is unknown connections. Who is to say that the parts are not connected by the transmission of unknown Q-rays and their detection by unrecognizable Q-detectors? One can only offer affidavits from the manufacturer testifying to an ignorance of Q-technology and, in any event, no such intent.

Evidently it is impossible to rule out conclusively the possibility of connections. The proper point of view to take, however, is that it is precisely the wonder and glory of the device that it impels one to doubt these assurances from one's own eyes and hands, professional magicians, and technical experts of all kinds. Suffice it to say that there are no connections that suspicious lay people or experts of broad erudition and unimpeachable integrity can discern. If you find yourself questioning this, then you have grasped the mystery of the atomic world.

Two of the three parts of the device (A and B) function as detectors. Each detector has a switch that can be set in one of three positions (1, 2, and 3) and a red and a green light bulb (Fig. 1).

When a detector is set off it flashes either its red light or its green. It does this no matter how its switch is set, though whether it flashes red or green may well depend on the setting. The only purpose of the lights is to communicate information to us; marks on a ribbon of tape would serve as well. I mention this only to emphasize that the unconnectedness of the parts prohibits a mechanism in either detector that might modify its behavior according to the color that may have flashed at the other.

The third and last part of the device is a box (C) placed between the detectors. Whenever a button on the box is pushed, shortly thereafter two particles emerge, moving off in opposite directions toward the two detectors (Fig. 2). Each detector flashes either red or green whenever a particle reaches it. Thus within a second or two of every push of the button, each detector flashes one or the other of its two colored lights.

Because there are no connections between parts of the device, the link between pressing the button on the box and the subsequent flashing of the detectors can be provided only by the passage of the particles from the box to the detectors. This passage could be confirmed by subsidiary detectors between the box and the main detectors A and B, which can be designed so as not to alter the functioning of the device. Additional instruments

Fig. 1. A detector. Particles enter on the right. The red (R) and green (G) lights are on the left. The switch is set to 1.

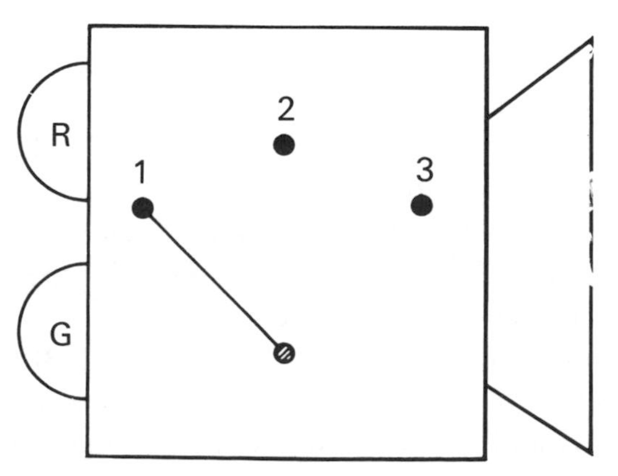

or shields could also be used to confirm the lack of other communication between the box and the two detectors or between the detectors themselves (Fig. 3).

The device is operated repeatedly in the following way. The switch on each detector is set at random to one of its three possible positions, giving nine equally likely settings for the pair of detectors: 11, 12, 13, 21, 22, 23, 31, 32, and 33. The button on the box is then pushed, and somewhat later each detector flashes one of its lights. The flashing of the detectors need not be simultaneous. By changing the distance between the box and the detectors we can arrange that either flashes first. We can also let the switches be given their random settings either before or after the particles leave the box. One could even arrange for the switch on B not to be set until after A had flashed (but, of course, before B flashed).

After both detectors have flashed their lights, the settings of the switches and the colors that flashed are recorded, using the following notation: 31 GR means that detector A was set to 3 and flashed green, while B was set to 1 and flashed red; 12 RR describes a run in which A was at 1, B at 2, and both flashed red; 22 RG describes a run in which both detectors were set to 2, A flashed red and B flashed green; and so on. A typical fragment from a record of many runs is shown in Fig. 4.

The accumulated data have a random character, but, like data collected in many tossings of a coin, they reveal certain unmistakable features when enormously many runs are examined. The statistical character of the data should not be a source of concern or suspicion. Blaming the behavior of the device on repeated, systematic, and reproducible accidents, is to offer an

Fig. 2. The complete device. A and B are the two detectors. C is the box from which the two particles emerge.

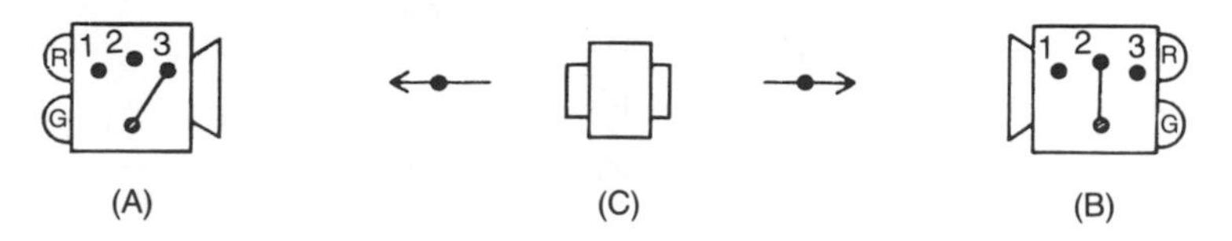

explanation even more astonishing than the conundrum it is invoked to dispel.

The data accumulated over millions (or, if you prefer, billions or trillions) of runs can be summarized by distinguishing two cases:

Case a. In those runs in which each switch ends up with the same setting (11, 22, or 33) both detectors always flash the same color. RR and GG occur in a random pattern with equal frequency; RG and GR never occur.

Case b. In the remaining runs, those in which the switches end up with different settings (12, 13, 21, 23, 31, or 32), both detectors flash the same color only a quarter of the time (RR and GG occurring randomly with equal frequency); the other three quarters of the time the detectors flash different colors (RG and GR occurring randomly with equal frequency).

These results are subject to the fluctuations accompanying any statistical predictions, but, as in the case of a coin-tossing

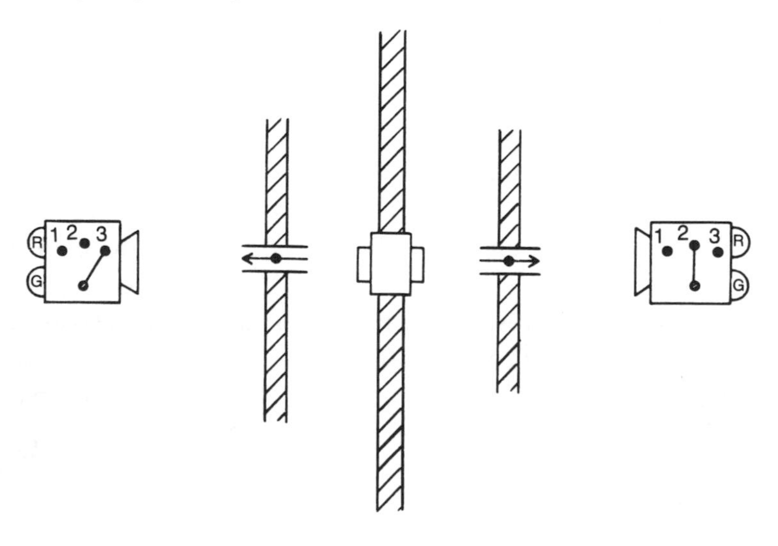

Fig. 3. Possible refinement of the device. The box is embedded in a wall that cuts off one detector from the other. Subsidiary detectors confirm the passage of the particles to the main detectors.

experiment, the observed ratios will differ less and less from those predicted, as the number of runs becomes larger and larger.

This is all it is necessary to know about how the device operates. The particular fractions $\frac{1}{4}$ and $\frac{3}{4}$ arising in Case b are of critical importance. If the smaller of the two were $\frac{1}{3}$ or more (and the larger $\frac{2}{3}$ or less) there would be nothing wonderful about the device. To produce the conundrum it is necessary to run the experiment sufficiently many times to establish with overwhelming probability that the observed frequencies (which will be close to 25% and 75%) are not chance fluctuations away from expected frequencies of $33\frac{1}{3}$% and $66\frac{2}{3}$%. (A million runs is more than enough for this purpose.)

These statistics may seem harmless enough, but some scrutiny

Fig. 4. Fragment of a page of a volume from the set of notebooks recording a long series of runs.

```
          GG 22GG 1
       2RR 31RG 13RG 22GG 22R
    R 21GR 32RG 11GG 32GR 33GG 2
  22GG 11RR 11GG 23GG 12RR 32GR 11GG
G 12RG 13RG 33GG 21RG 13GR 31RR 32GR
GR 13GR 21RG 33RR 13GR 11RR 11GG 13RG 31
2GG 32GR 33GG 21GR 21GG 33RR 23RG 21GG 21R
13GR 11GG 32GG 31GR 32RG 33RR 13RR 13RG 12R
11GG 31RG 33RR 12RG 21GR 11GG 22GG 33GG 23G
11RR 22RR 12RG 22GG 23GR 12GR 33GG 31GG 13G
13GR 21RR 33RR 33RR 13RG 23RG 33GG 32RR 12R
3RR 32RG 11RR 11RR 11RR 32RG 12RG 21RG 11G
RG 23RR 21RG 33RR 13GR 12GR 23RG 21RR 32
R 21GR 12RR 31GR 12RG 13GR 13RG 22RR 1
  23GR 11RR 12RR 33RR 21RG 13GR 21RR
    R 12RR 23GG 13RG 21RG 11GG 1
      2RG 32RG 32GR 11GG 22R
            31RG 2
```

reveals them to be as surprising as anything seen in a magic show, and leads to similar suspicions of hidden wires, mirrors, or confederates under the floor. We begin by seeking to explain why the detectors invariably flash the same colors when the switches are in the same positions (Case a). There would be any number of ways to arrange this were the detectors connected, but they are not. Nothing in the construction of either detector is designed to allow its functioning to be affected in any way by the setting of the switch on the other, or by the color of the light flashed by the other.

Given the unconnectedness of the detectors, there is one (and, I would think, only one) extremely simple way to explain the behavior in Case a. We need only suppose that some property of each particle (such as its speed, size, or shape) determines the color its detector will flash for each of the three switch positions. What that property happens to be is of no consequence; we require only that the various states or conditions of each particle can be divided into eight types; RRR, RRG, RGR, RGG, GRR, GRG, GGR, and GGG. A particle whose state is of type RGG, for example, will always cause its detector to flash red for setting 1 of the switch, green for setting 2, and green for setting 3; a particle in a state of type GGG will cause its detector to flash green for any setting of the switch; and so on. The eight types of states encompass all possible cases. The detector is sensitive to the state of the particle and responds accordingly; putting it another way, a particle can be regarded as carrying a specific set of flashing instructions to its detector, depending on which of the eight states the particle is in.

The absence of RG or GR when the two switches have the same settings can then be simply explained by assuming that the two particles produced in a given run are both produced in the same state; i.e., they carry identical instruction sets. Thus if both particles in a run are produced in states of type RRG, then both detectors will flash red if both switches are set to 1 or 2, and both will flash green if both switches are set to 3. The detectors flash the same colors when the switches have the same settings because the particles carry the same instructions.

This hypothesis is the obvious way to account for what happens in Case a. I cannot prove that it is the only way, but I challenge the reader, given the lack of connections between the detectors, to suggest any other.

The apparent inevitability of this explanation for the perfect correlations in case a forms the basis for the conundrum posed by the device. For the explanation is quite incompatible with what happens in Case b.

If the hypothesis of instruction sets were correct, then both particles in any given run would have to carry identical instruction sets whether or not the switches on the detectors were set the same. At the moment the particles are produced there is no way to know how the switches are going to be set. For one thing, there is no communication between the detectors and the particle-emitting box, but in any event the switches need not be set to their random positions until after the particles have gone off in opposite directions from the box. To ensure that the detectors invariably flash the same color every time the switches end up with the same settings, the particles leaving the box in each run must carry the same instructions even in those runs (Case b) in which the switches end up with different settings.

Let us now consider the totality of all Case b runs. In none of them do we ever learn what the full instruction sets were, since the data reveal only the colors assigned to two of the three settings. (The Case a runs are even less informative). Nevertheless we can draw some nontrivial conclusions by examining the implications of each of the eight possible instruction sets for those runs in which the switches end up with different settings. Suppose, for example, that both particles carry the instruction set RRG. Then out of the six possible Case b settings, 12 and 21 will result in both detectors flashing the same color (red), and the remaining four settings, 13, 31, 23, and 32, will result in one red flash and one green. Thus both detectors will flash the same color for two of the six possible Case b settings. Since the switch settings are completely random, the various Case b settings occur with equal frequency. Both detectors will therefore flash the same color in a third of those Case b runs in which the particles carry the instruction sets RRG.

The same is true for Case b runs where the instruction set is RGR, GRR, GGR, GRG, or RGG, since the conclusion rests only on the fact that one color appears in the instruction set once and the other color, twice. In a third of the Case b runs in which the particles carry any of these instruction sets, the detector will flash the same color.

The only remaining instruction sets are RRR and GGG; for these sets both detectors will evidently flash the same color in every Case b run.

Thus, regardless of how the instruction sets are distributed among the different runs, in the Case b runs *both detectors must flash the same color at least a third of the time.* (This is a bare minimum; the same color will flash more than a third of the time, unless the instruction sets RRR and GGG never occur.) As emphasized earlier, however, when the device actually operates the same color is flashed only a quarter of the time in the Case b runs.

Thus the observed facts in Case b are incompatible with the only apparent explanation of the observed facts in Case a, leaving us with the profound problem of how else to account for the behavior in both cases. This is the conundrum posed by the device, for there is no other obvious explanation of why the same color always flashes when the switches are set the same. It would appear that there must, after all, be connections between the detectors – connections of no known description which serve no purpose other than relieving us of the task of accounting for the behavior of the device in their absence.

I shall not pursue this line of thought, since my aim is only to state the conundrum of the device, not to resolve it. The lecture demonstration is over. I shall only add a few remarks on the device as a parable.

One of the historic exchanges between Einstein and Bohr,[3,4]

[3] A. Einstein, B. Podolsky, and N. Rosen, *Physical Review*, **47**, 777 (1935).
[4] N. Bohr, *Physical Review*, **48**, 696 (1935).

which found its surprising denouement in the work of J. S. Bell nearly three decades later,[5] can be stated quite clearly in terms of the device. I stress that the transcription into the context of the device is only to simplify the particular physical arrangement used to raise the issues. The device is a direct descendant of the rather more intricate but conceptually similar *gedanken* experiment proposed in 1935 by Einstein, Podolsky, and Rosen. We are still talking physics, not descending to the level of analogy.

The Einstein-Podolsky-Rosen experiment amounts to running the device under restricted conditions in which both switches are required to have the same setting (Case a). Einstein would argue (as was argued above) that the perfect correlations in each run (RR or GG but never RG or GR) can be explained only if instruction sets exist, each particle in a run carrying the same instructions. In the Einstein-Podolsky-Rosen version of the argument the analogue of Case b was not evident, and its fatal implications for the hypothesis of instruction sets went unnoticed until Bell's paper.

The *gedanken* experiment was designed to challenge the prevailing interpretation of the quantum theory, which emphatically denied the existence of instruction sets, insisting that certain physical properties (said to be complementary) had no meaning independent of the experimental procedure by which they were measured. Such measurements, far from revealing the value of a preexisting property, had to be regarded as an inseparable part of the very attribute they were designed to measure. Properties of this kind have no independent reality outside the context of a specific experiment arranged to observe them: the moon is *not* there when nobody looks.

In the case of my device, three such properties are involved for each particle. We can call them the 1-color, 2-color, and 3-color of the particle. The n-color of a particle is red if a detector with its switch set to n flashes red when the particle arrives. The three n-colors of a particle are complementary properties. The switch on a detector can be set to only one of the three positions, and the experimental arrangements for measuring the 1-, 2-, or 3-color of

[5] J. S. Bell, *Physics*, I, 195 (1964).

a particle are mutually exclusive. (We may assume, to make this point quite firm, that the particle is destroyed by the act of triggering the detector, which is, in fact, the case in many recent experiments probing the principles that underly the device.)

To assume that instruction sets exist at all, is to assume that a particle has a definite 1-, 2-, and 3-color. Whether or not all three colors are known or knowable is not the point; the mere assumption that all three have values violates a fundamental quantum-theoretic dogma.

No basis for challenging this dogma is evident when only a single particle and detector are considered. The ingenuity of Einstein, Podolsky, and Rosen lay in discovering a situation involving a *pair* of particles and detectors, where the quantum dogma continued to deny the existence of 1-, 2-, and 3-colors, while, at the same time, quantum theory predicted correlations (RR and GG but never RG or GR) that seemed to require their existence.

Einstein concluded that, if the quantum theory were correct, i.e. if the correlations were, as predicted, perfect, then the dogma on the nonexistence of complementary properties – essentially Bohr's doctrine of complementarity – had to be rejected.

Pauli's attitude toward this in his letter to Born is typical of the position taken by many physicists: since there is no known way to determine all three *n*-colors of a particle, why waste your time arguing about whether or not they exist? To deny their existence has a certain powerful economy – why encumber the theory with inaccessible entities? More importantly, the denial is supported by the formal structure of the quantum theory which completely fails to allow for any consideration of the simultaneous 1-, 2-, and 3-colors of a particle. Einstein preferred to conclude that all three *n*-colors did exist, and that the quantum theory was incomplete. I suspect that many physicists, though not challenging the completeness of the quantum theory, managed to live with the Einstein-Podolsky-Rosen argument by observing that though there was no way to establish the existence of all three *n*-colors, there was also no way to establish their nonexistence. Let the angels sit, even if they can't be counted.

Bell changed all this, by bringing into consideration the Case b runs, and pointing out that the quantitative numerical predictions of the quantum theory ($\frac{1}{4}$ vs. $\frac{1}{3}$) unambiguously ruled out the existence of all three *n*-colors. Experiments done since Bell's paper confirm the quantum-theoretic predictions.[6] Einstein's attack, were he to maintain it today, would be more than an attack on the metaphysical underpinnings of the quantum theory – more, even, than an attack on the quantitative numerical predictions of the quantum theory. Einstein's position now appears to be contradicted by nature itself. The device behaves as it behaves, and no mention of wave-functions, reduction hypotheses, measurement theory, superposition principles, wave–particle duality, incompatible observables, complementarity, or the uncertainty principle, is needed to bring home its peculiarity. It is not the Copenhagen interpretation of quantum mechanics that is strange, but the world itself.

As far as I can tell, physicists live with the existence of the device by implicitly (or even explicitly) denying the absence of connections between its pieces. References are made to the "wholeness" of nature: particles, detectors, and box can be considered only in their totality; the triggering and flashing of detector A cannot be considered in isolation from the triggering and flashing of detector B – both are part of a single indivisible process. This attitude is sometimes tinged with Eastern mysticism, sometimes with Western know-nothingism, but, common to either point of view, as well as to the less trivial but considerably more obscure position of Bohr, is the sense that strange connections are there. The connections are strange because they play no explicit role in the theory: they are associated with no particles or fields and cannot be used to send any kinds of signals.

[6] Theoretical and experimental aspects of the subject are reviewed by J. F. Clauser and A. Shimony, *Reports on Progress in Physics*, **41**, 1991 (1978). For a less technical survey see B. d'Espagnat, *Scientific American*, **240**, 5 (November 1979): 158.

They are there for one and only one reason: to relieve the perplexity engendered by the insistence that there are no connections.

Whether or not this is a satisfactory state of affairs is, I suspect, a question better addressed by philosophers than by physicists.

I conclude with the recipe for making the device, which, I emphasize again, can be ignored:

> The device exploits Bohm's version[7] of the Einstein, Podolsky, Rosen experiment. The two particles emerging from the box are spin $\frac{1}{2}$ particles in the singlet state. The two detectors contain Stern–Gerlach magnets, and the three switch positions determine whether the orientations of the magnets are vertical or at $\pm 120°$ to the vertical in the plane perpendicular to the line of flight of the particles. When the switches have the same settings the magnets have the same orientation. One detector flashes red or green according to whether the measured spin is along or opposite to the field; the other uses the opposite color convention. Thus when the same colors flash the measured spin components are different.
>
> It is a well-known elementary result that, when the orientations of the magnets differ by an angle θ, then the probability of spin measurements on each particle yielding opposite values is $\cos^2(\theta/2)$. This probability is unity when $\theta = 0$ (Case a) and $\frac{1}{4}$ when $\theta = \pm 120°$ (Case b).
>
> If the subsidiary detectors verifying the passage of the particles from the box to the magnets are entirely non-magnetic they will not interfere with this behavior.

[7] D. Bohm, *Quantum Theory* (Englewood Cliffs, N.J.: Prentice-Hall, 1951), pp. 614–19.

11

Can you help your team tonight by watching on TV? More experimental metaphysics from Einstein, Podolsky, and Rosen

A few years ago I described[1] a simple device that reveals in a very elementary way the extremely perplexing character the data from the Bohm–Einstein–Podolsky–Rosen experiment assumes in the light of the analysis of J. S. Bell. There is a second, closely related form of that *gedanken* demonstration,[2] which I would like to examine for several reasons.

1. It is simpler: there are only two (not three) settings for each switch.
2. The *gedanken* data resemble more closely the data collected in actual realizations of the device.
3. None of the possible switch settings produce the perfect correlations found in the first version of the *gedanken* demonstration, where the lights *always* flash the same color when the switches have the same setting. Since absolutely perfect correlations are never found in the imperfect experiments we contend with in the real world, an argument that eliminates this feature of the ideal *gedanken* data can be applied to real data from real experiments. (If you believe, however, along with virtually all physicists, that the

[1] N. David Mermin, *Journal of Philosophy*, **78**, 397 (1981), reprinted as the preceeding essay. An only slightly more technical but significantly more graceful version recently appeared in *Physics Today*, April, 1985.

[2] What follows is my attempt to simplify some reformulations of EPR and Bell by Henry Stapp (for example, *Am. J. Phys.* **53**, 306 (1985)), but the interpretation I give differs from his, and any foolishness in what follows is entirely my own.

quantum theory gives the correct ideal limiting description of all phenomena to which it can be applied, then this is not so important a consideration.)

4. Because the ideal perfect correlations are absent from this version of the *gedanken* demonstration, one is no longer impelled to assert the existence of impossible instruction sets. To establish that the new data nevertheless remain peculiar, it is necessary to take a different line of attack, which has again intriguing philosophical implications, but of a rather different character.[3]

The modified demonstration

In the modified *gedanken* demonstration there are only two switch settings (1 and 2) at each detector. Otherwise the set-up is unchanged: there are two detectors (A and B) and a source (C), and the result of each run is the flashing of a red or green light. If one had actually built such a device according to the quantum mechanical prescription, it could be transformed to run in this modified mode simply by readjusting the angle through which certain internal parts of each detector turned as the switch settings were changed.[4]

In its new mode of operation, the device produces the following data:

(i) When the experiment is run with both switches set to 2 (22 runs) the lights flash the same color only 15% of the time; in 85% of the 22 runs different colors flash.

[3] There are more orthodox ways of extracting the peculiar character of this data. The route I take here requires fewer formal probabilistic excursions, and leads to a rather different philosophical point, though I suspect a careful analysis of the use of probability distributions in the conventional arguments might uncover something quite similar.

[4] Physicists might note that if setting #1 at detector A corresponds to measuring the vertical spin component, then setting #2 at A measures the component at 90° to the vertical setting #1 at B, 45° to the vertical, and setting #2 at B, −45° to the vertical, all four directions lying in the same plane. (In the earlier version the three switch settings at either detector corresponded to 0°, 120°, and −120°.) The fraction 85% is just $\cos^2(22.5°) = \frac{1}{4}(2 + \sqrt{2})$.

(ii) When the experiment is run with any of the other three possible switch settings (11, 12, or 21 runs) then the lights flash the same color 85% of the time; in only 15% of these runs do different colors flash.

As in the earlier version of the *gedanken* experiment, RR and GG are equally likely when the lights do flash the same color, and RG and GR are equally likely when different colors flash. Also as earlier the pattern of colors observed at any single detector is entirely random. There is no way to infer from the data at one detector how the switch was set at the other. Regardless of what is going on at detector B, the data for a great many runs at detector A is simply a random string of R's and G's, that might look like this:

Typical data at detector A

A: R G R R G G R G R R R G G G R G R R R G R G R G...

The choice of switch settings only affects the *relation* between the colors flashed at *both* detectors. If, for example, the above data had been obtained at detector A when its switch was set to 2, and in all those runs the switch at B had also been set to 2, then, as noted above, the color flashed at B would have agreed with that flashed at A in only 15% of the runs, and the lights flashed at both detectors together might thus have looked like this:[5]

Data from a series of 22 runs

A2: R G R R G G R G R R R G G G R G R R R G R G R G...

B2: G R G G G R G R G G R G R R G R G G R R R R G R...

Although the list of colors flashed at either detector remains quite random, the color flashed at B is highly (negatively) correlated with the color flashed at A. In the overwhelming majority (85%) of the runs the detectors flash different colors. Only in a few (15%) of the runs do the detectors flash the same color.

[5] The numbers after A and B denote the fixed setting of their switches throughout the sequence of runs. In contrast to some earlier version of the *gedanken* demonstration, we now try out various fixed switch settings, rather than randomly resetting the switches after each run.

On the other hand, for any of the other switch settings (take 21 as an example) the comparative data would have looked something like this:

Data from a series of 21 runs

A2: R G R R G G R G R R R G G G R G R R R G R G R G...
B1: R G R G G G R R R R R G G R G G R R R G G G R G...

Again we have two lists of colors, each entirely random, but they now agree with each other in 85% of the runs, disagreeing in only 15%.

There are various ways to run the modified *gedanken* demonstration, but let me focus on the following procedure, which it seems to me makes a rather striking contribution to Abner Shimony's field of experimental metaphysics. Suppose we do a long series of runs in each of which both switches are set to 1:

Data from a series of 11 runs

A1: R G G G R G R G R R G R R G R G R R R R G R G G...
B1: R G G R R G G R R R G R R G R G R G R R G G G G...

About 85% of these 11 runs will produce the same colors, and 15%, different. Now because there are no connections of any kind between the detectors at A and B, it seems clear that whatever happens at A cannot in any way depend on how the switch was set at B, and vice-versa. Let us elevate this common sense remark into a principle, which I shall call the Baseball Principle. Before examining the implications of the *gedanken* demonstration for the Baseball Principle, let us discuss it in the context from which its name derives, where it assumes (at least for me) an especially vivid character.

The Baseball Principle

I'm a New York Mets fan, and when they play a crucial game I feel I should watch on television. Why? Not just to find out what's going on. Somewhere deep inside me, I feel that my watching the game makes a difference – that the Mets are more likely to win if I'm following things than if I'm not. How can I say such a thing? Do I think, for example, that by offering up little prayers at crucial moments I can induce a very gentle divine intervention that will produce the minute change in trajectory of bat or ball that makes the difference between a hit or an out? Of course not! My feeling is completely irrational. If you insisted that I calm down and think about it, I'd have to admit that the outcome of the game doesn't depend in the least on whether I watch it or not. What I do or don't do in Ithaca, New York, can have no effect on what the Mets do or don't do in Flushing, New York. This is the Baseball Principle.

Now a pedant comes along and says, "What do you really mean by that Baseball Principle?" And then, being a pedant, he tells me what I really mean. What I really mean is this: If we examined a great many Mets games and divided them up into those I watched at least part of on TV and those I didn't watch at all, and if my decision to watch or not was entirely independent of anything I knew about the game – made, for example, by tossing a coin – then we would find that the Mets were no more or less successful in those games I watched than in those games I didn't.

Now I reply, "That's very nice, but I mean something much simpler. I mean that in each individual game, it doesn't make any difference whether I watch it or not. Tonight, for example, whatever the Mets do, will be exactly the same, whether or not I end up watching the game."

"C'mon," says the pedant, "that's silly. Either you watch the game or you don't. You can't say that what happens in the game in the case that didn't happen is the same as what happens in the case that did, because there's no way to check. What didn't happen *didn't happen*."

I say to the pedant: "Who's being silly here? Are you trying to tell me that it *does* make a difference in tonight's game whether I watch it or not?"

"No," says the pedant, "I'm saying that your statement that it doesn't make any difference whether or not you watch an individual game, can only be viewed as a very convenient construct to summarize the more complex statistical statement about correlations between watching and winning over many games. All of its statistical implications are correct, but it has no meaning when applied to an individual game, because there is no way to verify it in the case of the individual game, which you cannot both watch and not watch."

But is it *wrong* to apply the Baseball Principle to an individual game?

The Strong Baseball Principle

Let us call the claim that the Baseball Principle applies to each individual game the Strong Baseball Principle. The Strong Baseball Principle insists that the outcome of any particular game doesn't depend on what I do with my television set – that whatever it is that happens *tonight* in Shea Stadium, will happen in exactly the same way, whether or not I'm watching on TV.

As a rational person, who is not superstitious, and does not believe in telepathy or the efficacy of prayer on the sporting scene, I am convinced of the Strong Baseball Principle. True, there is no way to verify it, since I cannot both watch and not watch tonight's game, and am therefore unable to compare how the game goes in both cases to make sure nothing changes. Nevertheless, deep in my heart, I do believe that because there is no mechanism connecting what I do with the TV at home to what happens in Shea Stadium, the outcome of tonight's game genuinely does not depend on whether I watch it or not: The Strong Baseball Principle. Try as you may to persuade me that the Strong Baseball Principle is meaningless, in my heart, I know it's right.

Remarkably, when run in the second mode, the *gedanken* demonstration provides us with a case in which if it really does make no difference whether or not I watch the game, then it is not only meaningless, but demonstrably *wrong* to assert this principle in the individual case. If the Baseball Principle is right for the device, then the Strong Baseball Principle must be wrong, not merely because it naively compares possibilities only one of which can be realized, but because it is directly contradicted by certain observed facts. Such an experimental refutation of the Strong Baseball Principle would have been impossible before the discovery of the quantum theory; you cannot get into trouble using the Strong Baseball Principle in classical physics, and it can, in fact, be a powerful conceptual tool.[6] I believe that those who take the view that an experimental refutation is of no interest since reasoning from the Strong Baseball Principle was impermissible all along, miss something of central importance for an understanding of the character of quantum phenomena.

The device and the Baseball Principle

We return from ball games to the device. There are no connections between the detectors or between the source and either detector. The Baseball Principle therefore applies, and asserts that what goes on at detector A does not depend on how the switch is set at detector B, and vice-versa. This is readily verified in the statistical sense insisted on by the pedant. Keep the switch at A set to 1. Do a great many runs with the switch at B set to 1. Then, keeping the switch at A at 1, do a second series of runs with the switch at B set to 2. Compare the data at A in the two cases. It will have exactly the same character – namely a featureless sequence of Rs and Gs like the series of heads and tails you get by repeatedly flipping a coin. There is nothing in the outcome at A to distinguish between the runs in which B was set to 1 or to 2.

But what about the Strong Baseball Principle? Given the lack of any connection between the detectors, can we not also assert that what goes on at one detector in any *individual* run of the experiment does not depend on how the switch is set at the other detector? Granted, there is no way to test this stronger assertion, but surely, for the same reason, there is also no way to refute it. But here, remarkably in my opinion, we have a case in which the

[6] In a deterministic world in which the future can be calculated from present conditions, the Strong Baseball Principle can be given an unambiguous meaning.

Strong Baseball Principle is directly contradicted by the data. Consider what happened when the device was run with both switches set to 1:

Actual data from a series of 11 runs

A1: R G G G R G R G R R G R R G R G R R R R G R G G...
B1: R G G R R G G R R R G R R G R G R G R R G G G G...

If there are really no connections between A and B, and no spooky actions at a distance, then what happens at detector A can't depend on how the switch is set at detector B (and vice-versa). The Strong Baseball Principle takes this to mean that in the first run of this sequence (in which both lights flashed R) the light at detector A would have flashed R even if the switch on detector B had been set to 2 instead of 1, and similarly, for every other run in the series, if B had been set to 2 nothing would have changed at A. In no individual run can the outcome at A depend on how the switch was set at B. (Compare this with "In no individual baseball game can the outcome at Shea Stadium depend on how the switch was set on my TV.")

Well if that's so, we can say something about what would have happened if the run had been 12 (A1 and B2) rather than 11 (A1 and B1) – namely the outcomes at A would have been exactly the same as before:[7]

The 11 runs and what the Strong Baseball Principle can say about what would have happened had they been 12 runs

B2: ?...
A1: R G G G R G R G R R G R R G R G R R R R G R G G...

A1: R G G G R G R G R R G R R G R G R R R R G R G G...
B1: R G G R R G G R R R G R R G R G R G R R G G G G...

Note that in this application of the Strong Baseball Principle we

[7] This does not imply determinism – indeed, I'm not convinced that what happens in a baseball game *is* deterministic; it simply says, in the baseball case, that whatever it is that does happen isn't going to depend on what a television set 300 miles away is doing.

make no commitment at all to what colors flashed at B in the case that didn't take place (with the switch at B set to 2) since, after all, that didn't happen. We merely assert that whatever might have taken place at B in that unrealized experiment, nothing would have turned out any differently at A.

We can also say the same thing about what would have happened at B, if we had set the switch differently at A. This gives us one more pair of rows:

The 11 runs and what the Strong Baseball Principle can say about what would have happened had they been 12 runs or what would have happened had they been 21 runs

B2: ? ...
A1: R G G G R G R G R R G R R G R G R R R R G R G G...

A1; R G G G R G R G R R G R R G R G R R R R G R G G...
B1: R G G R R G G R R R G R R G R G R G R R G G G G...

B1: R G G R R G G R R R G R R G R G R G R R G G G G...
A2: ? ...

Consider now what we have laid out here. The middle two (3rd and 4th) rows show what actually happened: both switches were set to 1, and the first run gave RR, the second, GG, the third GG, the fourth GR, etc. The top two rows (1st and 2nd) express the Strong Baseball Principle in the form that asserts that the outcome of *each individual* run at A does not depend on how the switch is set at B. The bottom two (5th and 6th) express it as an assertion that the outcome of each run at B does not depend on the switch setting at A.

Now what about the question marks? They appear in the top and bottom rows because those rows represent what would have happened at B and A had the switches there been other than what they actually were. Evidently *some* sequence of Rs and Gs would have been produced in either case,[8] but we have no way of telling

[8] At this moment in my talk there were cries of protest from the philosophers in the house. I was told that "If I were hungry I would eat a candy bar" does not imply the proposition "There exists a candy bar which is the one I would eat were I hungry" (the Candy Bar Principle). I affirmed my commitment to the Candy Bar Principle. I

which. Experience with the device, however, tells us some of the features these sequences would have had, if the runs had been 12 or 21 runs rather than the 11 runs that actually took place. An acceptable sequence of Rs and Gs for the first (B2) row, must agree with the sequence of Rs and Gs in the second (A1) row in about 85% of the positions, since that is the way 12 runs always work. Similarly a sequence of Rs and Gs replacing the question marks in the sixth row must agree in about 85% of the positions with the sequence in the fifth row, since that is what always happens in 21 runs. These considerations cut down on the number of ways of replacing question marks with Rs and Gs, but many different possibilities are still allowed.

A final application of the Strong Baseball Principle can be made to restrict these possibilities further. Suppose both switches had been set to 2 rather than 1. We can regard this 22 series of runs either as a modification of a 21 series (modified by changing the switch setting at B without changing anything at A) or as a modified 12 series (in which the switch was changed at A without anything having been done at B). We don't know, of course, what would have happened at B in the hypothetical 12 series (top row of question marks) or at A in the hypothetical 21 series (bottom row of question marks). The Strong Baseball Principle asserts that whatever series of Rs and Gs at A the question marks in the bottom row might stand for in the 21 run, that same series of Rs and Gs would also have happened at A in that series of runs, had the switch at B been set to 2 instead of 1 – i.e. had the runs been 22 runs instead. By the same token, whatever sequence of Rs and Gs the question marks in the top row represented for the results at B in a series of 12 runs, that same sequence would also have described the results at B had the runs been 22 runs.

This last application of the Strong Baseball Principle, by comparing hypothetical cases, has a different character than the

said I wanted to make a rather different point, but I think they all stopped listening then and there. I hope you will not stop reading here and now. If you insist on talking candy, I would suggest that a more accurately analogous proposition is "Either there exists a candy bar which is the one I would eat were I hungry or there does not."

first two, which compare a hypothetical case with the real one, and here it might more accurately be termed the Very Strong Baseball Principle. Returning to the sporting analogy, the Very Strong Baseball Principle applies when the game is, in fact, cancelled because of rain. I nevertheless maintain that had the game been played, it would have taken place in exactly the same way, whether or not I watched it. This last assertion, may elicit an even more violent objection from the pedant. Is it really reasonable to insist that something should happen in exactly the same way when conditions change very far away from it, when in actual fact it never happened at all?

But is it really any more reasonable, I hasten to add, to insist that such an assertion is impermissible? I maintain that if last night's game hadn't been rained out, it would have happened the same way whether or not I had watched it on television. Can you prove me wrong when I say this? Wouldn't most unsuperstitious people regard the proposition as true? Indeed, as uninterestingly true? To be sure, the pedant will translate it into a series of harmless statistical assertions, but is it really *wrong* to apply it to the individual case as well? The hallmark of the Strong Baseball Principle at work is that nagging conviction that only a pedant could object. For how can one possibly get into any trouble asserting relations between two things neither of which actually happened?

One can. It is worse than bad form; it is bad physics. For let's try it out. We have to replace the 1st row with some sequence of Rs and Gs and the 6th row with some other such sequence in such a way that the 1st and 2nd rows give the right statistics for 12 runs, the 5th and 6th, for 21 runs, and the 1st and 6th for 22 runs. We do not insist that any particular way of doing this is preferable to or any more deserving of some hypothetical reality than any other, but for the Strong Baseball Principle to survive, *some* among the various possibilities must be consistent with these statistics.

Now in 22 runs the colors disagree 85% of the time, so whatever goes into the 1st row has to disagree with whatever goes into the 6th in about 85% of the positions.

On the other hand the set of Rs and Gs in the top row can differ from that in the second row in only about 15% of the positions (since they must have the correlations appropriate to a series of 12 runs). The second row is the same as the third row (by the Strong Baseball Principle). The third row differs from the fourth row in only about 15% of the positions (since they give the data in a 11 run). The fourth row is the same as the fifth row (by the Strong Baseball Principle). And the fifth row can differ from the set of Rs and Gs appearing in the bottom row in only 15% of the positions (since those rows must have the correlations appropriate to a series of 21 runs).

A moment's reflection on the last paragraph is enough to reveal that whatever the sequence of Rs and Gs in the top row, it can differ from whatever sequence is in the bottom row, in at most about 15% + 15% + 15% = 45% of the positions. But according to the next to the last paragraph whatever is in the top row must differ from whatever is in the bottom row in about 85% of the positions. You can't have it both ways. Thus the (Very) Strong Baseball Principle is so restrictive as to rule out *every* possibility for the unrealized switch settings. Far from merely being meaningless nonsense, an application of the Strong Baseball Principle to the *gedanken* demonstration contradicts the observed facts.

In this demolition of the Strong Baseball Principle we did not interpret it as demanding the existence in some cosmic bookkeeping office of a list of data for the unperformed runs. We only took it to require that if the *actual* experiment consists of a long series of 11 runs, then among all the *possible* sets of data that *might* have been collected had the experiment instead consisted of 12, 21, or 22 runs, there should be *some* satisfying the condition that run by run what happens at one detector does not depend on how the switch is set at the other. If the Strong Baseball Principle is valid it should be possible to *imagine* sets of B2 and A2 data such that the B2 data produce the right statistics (85% same and 15% different) when combined with the actual A1 data, the A2 data produce the right statistics (85% same and 15% different) when combined with the actual B1 data, and the two sets of imagined data produce the right statistics (15% same and 85% different) for a 22 experiment.[9]

Since it is impossible to imagine *any* such sets of data then the Strong Baseball Principle has to be abandoned not because it is bad form, unjustifiable, or frivolous to argue from what might have happened but didn't, but because there are no conceivable sets of data for the cases that might have happened but didn't, which are consistent with the numerical constraints imposed by the known behavior of the device, when those constraints are further restricted by the Strong Baseball Principle.

This attack is inherently non-classical. If, in the best *gedanken* demonstration I could devise, the 85% and 15% had been replaced by 75% and 25%, then the argument would have collapsed. For instead of the top row being able to differ from the bottom by no more than 15% + 15% + 15% = 45%, which is manifestly less than the required 85%, it would only have been possible to bound the difference by 25% + 25% + 25%, which is just enough to provide the required 75%. Only by exploiting *quantum* correlations can one construct an 85%–15% *gedanken* demonstration. Any model of the device one might devise based on classical physics would necessarily result in 75%–25% or less extreme statistics, and the Strong Baseball Principle would be immune from this kind of refutation by physicists, no matter how dim a view of it philosophers took. I assert this with confidence because classical physics is local and deterministic and in a deterministic world the Strong Baseball Principle makes perfect sense as a manifestation of locality.

Going in the other direction, it is easy to invent fictitious *gedanken* demonstrations that produce data that refute the Strong Baseball Principle even more resoundingly than that of the

[9] In Candy Bar terms, the Strong Baseball Principle does not say that there exists a particular sequence of Rs and Gs which are the colors that would have flashed had a detector been set differently. It only says that among all the mutually exclusive and exhaustive possibilities for such sequences should be *some* that are consistent with the frequencies of flashings characteristic of the four different pairs of switch settings.

device. Consider, for example, a hypothetical device in which 85% and 15% were replaced by 100% and 0%, so that the lights always (not just most of the time) flashed the same color in 11, 12, and 21 runs, and never (not just infrequently) flashed the same colors in 22 runs. Then the argument refuting the Strong Baseball Principle would be even simpler. A 11 run would necessarily result in the same color (say R) at A and B. Suppose instead the switch at A had been set to 2. The Strong Baseball Principle would then assert that R would still have flashed at B, and since the same color always flashes in 12 runs, A would still have flashed R. By the same token B would still have flashed R had its switch been set to 2. Therefore, since the setting of the switch at one detector cannot affect what happens at the other, both would have flashed R if both had been set to 2. But when both are set to 2, both have to flash different colors.

No experiment is known that can provide this more compact refutation. Even quantum miracles can go only so far. The 85%–15% statistics are the most extreme I know how to extract from the quantum theory, and though they are strong enough to demolish the Strong Baseball Principle, the argument we went through is somewhat less direct than that available for the 100%–0% statistics.

It is a characteristic feature of all quantum conundrums that something has to have a non-vanishing probability of happening in two or more mutually exclusive ways for startling behavior to emerge. The viewpoints of quantum and classical physics are distinguished, more than anything else, by the impropriety in quantum physics of reasoning from an exhaustive enumeration of two or more such possibilities in cases that might have happened but didn't. We are startled when such reasoning fails, because as an analytical tool in classical physics and everyday life it is not only harmless but often quite fruitful. The most celebrated of all quantum conundra – how can there be a diffraction pattern when the electron had to go through one slit or the other? – is based on precisely this impropriety. It is just where there is room for some interplay between various unrealized possibilities, that one can look for the quantum world to perform for us the most magical of its tricks.

Therefore it is wrong to apply to individual runs of the experiment the principle that what happens at A does not depend on how the switch is set at B. Many people want to conclude from this that what happens at A *does* depend on how the switch is set at B, which is disquieting in view of the absence of any connections between the detectors. The conclusion can be avoided, if one renounces the Strong Baseball Principle, maintaining that indeed what happens at A does not depend on how the switch is set at B, but that this is only to be understood in its statistical sense, and most emphatically cannot be applied to individual runs of the experiment. To me this alternative conclusion is every bit as wonderful as the assertion of mysterious actions at a distance. I find it quite exquisite that, setting quantum metaphysics entirely aside, one can demonstrate directly from the data and the assumption that there are no mysterious actions at a distance, that there is no conceivable way consistently to apply the Baseball Principle to individual events.

12

Spooky actions at a distance: mysteries of the quantum theory

There was a time when the newspapers said that only twelve men understood the theory of relativity. I do not believe there ever was such a time. There might have been a time when only one man did, because he was the only guy who caught on, before he wrote his paper. But after people read the paper a lot of people understood the theory of relativity in some way or other, certainly more than twelve. On the other hand I think I can safely say that nobody understands quantum mechanics.

Richard P. Feynman[1]

I. Introduction: the quantum theory

The quantum theory was born in 1900, with the twentieth century, and future centuries will list it among our own's most remarkable achievements. Designed to account for the puzzling behavior of matter at the submicroscopic scale of individual atoms, the theory has enjoyed phenomenal success. It has accounted in a quantitative way for atomic phenomena with a numerical precision never before achieved in any field of science. When it was subsequently applied within the atom to the atomic nucleus – an object some hundred thousand times smaller – no further modification in the theory was needed. More recently, when applied within the particles making up the nucleus to describe their own internal structure – an investigation on a scale as much smaller than atoms as atoms are smaller than us – the quantum theory still shows no signs of requiring extension or revision. Contemporary technology, whether you approve of it or not, grew directly out of the new understanding of matter provided by the quantum theory, from nuclear power to the electronic devices that run your digital watch and the computer on which I am composing this essay.

But these formidable accomplishments by themselves do not account for the fascination, verging on awe, that the quantum theory inspires in thoughtful people who have learned the description of nature it provides. The basis for this reverent puzzlement is that the quantum theoretic description of physical reality is exquisitely strange and profoundly mysterious.

As noted by Richard Feynman, similar things have been said about the theory of relativity, which also revolutionized the way we think about the world, and also seems distinctly peculiar when first encountered. The mysteries of relativity fade, however, when one firmly recognizes that clocks do not measure some preexisting thing called "time," but that our concept of time is simply a convenient way to abstract the common behavior of all those objects we call "clocks." While such a distinction may sound like splitting hairs, it is remarkably liberating to realize that time in itself does not exist except as an abstraction to free us from having always to talk about this clock or that. The discovery that there is no time – only clocks – has deep and surprising consequences for many very simple things we tend to take for granted because of our almost instinctual conviction that time has a reality that transcends the behavior of clocks. Once one accepts this discovery and learns to recognize the many ways in which the erroneous belief in an absolute time infects our thinking and our language, the mysteries of relativity vanish.

The mysteries of the quantum theory are not so readily dispelled. To be sure, the theory requires us to abandon false prejudices just as strong and pervasive as the wrong ideas about time that relativity forces us to shed, but even after those misconceptions are discarded, a broad residue of mystery still remains.

The quantum theory was strange from the moment of its birth in 1900, when Max Planck discovered he could account precisely for certain features of the radiation in a hot oven by introducing into a routine calculation a step which he could neither justify nor explain. Only with this step would his computation yield the observed behavior. Einstein, in 1905, abstracting from Planck's unorthodox analysis the idea that radiant energy might have an unexpected and inexplicable lumpiness to it, was able to exploit the idea of light as a shower of particles (now called "photons") to resolve some puzzles connected with the way light knocks electrons out of a piece of metal.

Einstein's explanation was as baffling as Planck's, because it was well established that light consisted of continuous waves – not discrete particles. Worse still, although Einstein's explanation of the photoelectric effect relied on light behaving as a shower of particles, the properties of those particles depended critically on the features of the continuous waves his particulate description seemed to contradict.

Fundamental physics developed in this *ad hoc* apparently inconsistent fashion for about a quarter of a century. Niels Bohr succeeded in extending such arguments (without explaining how they could possibly make sense) to account with remarkable precision for many of the intricate and detailed features of the light emitted and absorbed by simple atoms like hydrogen. Louis de Broglie observed that additional explanatory power was achieved by extending the confusion and recognizing that not only should light waves be viewed as showers of particles but that electrons, entities very well established as particles, might also be fruitfully regarded as waves.

In 1925 Werner Heisenberg and Erwin Schrödinger independently constructed a new theory – the modern form of the *quantum theory* (also known as *quantum mechanics* or, less frequently these days, *wave mechanics*) – that set in a single coherent scheme all the remarkably successful but apparently self-contradictory guesses on which the preceding twenty-five years of physics had relied. Quantum mechanics provided the unambiguous calculational method that has underlain all the explosive growth and flowering that physical science has enjoyed from 1925 right up to the present moment.

But the quantum theory remains deeply mysterious. It is no harder to use than any other branch of physics, and thousands – indeed, hundreds of thousands – of people have mastered its computational intricacies since it was first put forth. It is capable, in principle, of predicting the outcome of any experiment one can describe precisely enough to apply the mathematical apparatus of the theory. What makes it mysterious is that in general the quantum theory refuses to offer any picture of what is actually going on out there. If you ask it a question of the form "If I do this then what will I find if I measure that?" it will give you the answer. But if you ask, "Explain why I get that answer?" or, in certain especially intriguing examples that will occupy us here, "How can that possibly be the answer?" it is silent.

This strange quality of the theory was well appreciated by its founders and has been remarked upon by many knowledgeable people ever since. The following comments on quantum mechanics, by – or attributed to – some of the greatest physicists of the century, should convey something of the awe in which the theory is held. I began this essay with a statement by Feynman, who was too young to participate in the construction and exuberant application of the theory that took place in 1925 and the decade that followed, but who is probably the most gifted practitioner of quantum physics in the first generation to have grown up with it.

Heisenberg, describing the turbulent period in the mid-twenties when he and Bohr were fighting their way through to a coherent statement of the new physics, said of their struggles some thirty years later:

> An intensive study of all questions concerning the interpretation of quantum theory in Copenhagen finally led to a complete and, as many physicists believe, satisfactory clarification of the situation. But it was not a solution which one could easily accept. I remember discussions with Bohr

> which went through many hours till very late at night and ended almost in despair; and when at the end of the discussion I went alone for a walk in the neighboring park I repeated to myself again and again the question: Can nature possibly be as absurd as it seemed to us in these atomic experiments?[2]

Bohr, it should be noted, never said anything as dramatic in his own writings on the subject, but he was apparently less inhibited in private conversation:

> [Bohr] also said once that if somebody said, "Well this is clear," then he could say "Well, I think that if a man says it is completely clear to him these days, then he has not understood really the subject." He said once, I think he said it in German, "If you do not get *schwindlig* [dizzy] sometimes when you think about these things then you have not really understood it."[3]

> When asked whether the algorithm of quantum mechanics could be considered as somehow mirroring an underlying quantum world, Bohr would answer, "There is no quantum world. There is only an abstract quantum physical description. It is wrong to think that the task of physics is to find out how nature is. Physics concerns what we can say about nature."[4]

This surprising view (which Bohr himself never put in writing) that physics does not describe the world, but only what we know about the world, appears often in the later writings of Heisenberg:

> The conception of the objective reality of the elementary particles has thus evaporated in a curious way, not into the fog of some new, obscure, or not yet understood reality concept, but into the transparent clarity of a mathematics that represents no longer the behavior of the elementary particles but rather our knowledge of this behavior.[5]

> All the opponents of the Copenhagen interpretation do agree on one point. It would, in their view, be desirable to return to the reality concept of classical physics. ... They would prefer to come back to the idea of an objective real world whose smallest parts exist objectively in the same sense as stones or trees exist, independently of whether or not we observe them. This, however, is impossible, or at least not

entirely possible because of the nature of the atomic phenomena.[6]

Quite generally there is no way of describing what happens between two consecutive observations. It is of course tempting to say that the electron must have been somewhere between the two observations and that therefore the electron must have described some kind of path or orbit even if it may be impossible to know which path. This would be a reasonable argument in classical physics. But in quantum theory it would be a misuse of the language which ... cannot be justified.[7]

In recent years various distorted versions of this striking but now deeply orthodox and broadly accepted point of view have received a certain amount of popular attention. A recent editorial blurb in *Discover* remarks,

Physicists are wondering whether a tree – or anything else – must be observed before it really exists.[8]

Less accurately and even more sensationally, a similar blurb in the ordinarily staid *Scientific American* maintains:

The doctrine that the world is made up of objects whose existence is independent of human consciousness turns out to be in conflict with quantum mechanics and with facts established by experiment.[9]

Extravagant statements like these, especially those of the founding fathers, were not well received in all quarters. Einstein, in particular, though he had, with Planck, started the ball rolling, was already deeply disapproving in 1928:

The Heisenberg–Bohr tranquilizing philosophy – or religion? – is so delicately contrived that, for the time being, it provides a gentle pillow for the true believer from which he cannot very easily be aroused.[10]

This view persisted to the end of Einstein's life. In his study "Einstein and the Quantum Theory" Abraham Pais recalls conversations he had with him in Princeton:

We often discussed his notions on objective reality. I recall that during one walk Einstein suddenly stopped, turned to me and asked whether I really believed that the moon exists only when I look at it.[11]

What is at issue here – what lies behind all this dramatic talk – is the question of whether it is permissible to assign values to, or even speak at all of, physical properties whose values one has not actually ascertained by a measurement. The original reason for this ontological shyness is not mysterious. It has to do with the atomicity of everything at the ultramicroscopic level. To measure the position of something, for example, one has to observe where it is. We arc used to thinking of observation as a gentle process by which we acquire information without significantly disrupting what we observe, but observing an electron does not work like this. To see where it is, for example, we must shine light at it and collect some of the light that bounces off. (This is what "seeing" always conists of.) Now light is not a continuously alterable wave but a collection of discrete particles – Einstein's photons. Individual photons are minute things, with a negligible ability to disrupt the books or paintings we bounce them off when we turn on the light to look. But an electron is also very small. When we try to see where it is by illuminating it, we are, from the electron's point of view, bombarding it with a shower of photons, any of which can give it a violent kick.

Properties, at the atomic level, are unavoidably and appreciably disrupted by the very act of collecting information about them. How this must happen was analyzed in great detail by Bohr and especially by Heisenberg. They found that such disruptions could be reduced by well-designed methods of gathering information, but not entirely eliminated without at the same time losing some of the information one was trying to collect. The minimal unavoidable disruption is described by Heisenberg's celebrated uncertainty principle (occasionally also called the "indeterminacy principle").

The uncertainty principle says, for example, that by being sufficiently careful in your measurement of the position of an electron you can, in fact, determine that position to whatever precision you require, but you must pay a price. The unavoidable disruption attendant upon measuring its position will disturb, in an uncontrollable manner, the velocity of the electron. The more accurately you determine the position, the greater the uncertainty you introduce in the velocity. It works the other way, too. There are ways of measuring the velocity to any precision you wish, but because velocity measurements are also disruptive, the more accurately you measure the velocity, the greater uncertainty you introduce into where the electron is.

This is perhaps the single most striking difference between behavior at the macroscopic level at which we operate and the microscopic level of individual atomic particles. If this kind of behavior were scaled up to our level, then it would not, for example, be possible for a police helicopter to radio a patrol car that somebody was going at 75 miles per hour on route 17 at the intersection with exit 106. On the contrary, if the helicopter had ascertained that the speed of the car was exactly 75 miles per hour, then it would be incapable of specifying where the automobile was at all. If it contented itself with establishing that the speed was between 73 and 77 miles per hour, then it might only be able to say that the car was somewhere between exit 96 and exit 116. To be able to announce that the car was somewhere between exits 105 and 107, it might have to settle for the information that the speed was somewhere between 60 and 90 miles per hour.

Evidently this is a very unfamiliar kind of limitation, and in the case of speed control we would all know what to do: improve the police radar – give them better equipment so they can refine both their position and their velocity measurements together. At the atomic level this is impossible. The uncertainty principle expresses an inherent limitation to the precision with which both position and velocity can be measured at once. No improvement is possible.

The following question now arises. If it is indeed absolutely impossible to measure both the position and velocity of an electron with arbitrary precision, does it make sense to say that an electron has both a position and a velocity at all? If you think it does, then you are committed to the view that electrons, like automobiles, exist in definite places while moving with definite speeds, but there just happen to be no methods available to learn both the location and the speed of an electron. Both properties are there, but we cannot learn the values of both together.

There is a second point of view, and it is the position taken by all practitioners of the quantum theory. One maintains that it was foolish of us ever to regard electrons as the kinds of things that could have at once both a position and a velocity. After all, the ideas of position and velocity that we have evolved are based on our experience of objects about our own size, or not very much smaller. As we managed to extend our ability to study objects down to smaller and smaller scales, microscopic and then submicroscopic, we should never have expected that all the concepts we were able to apply to larger objects should be applicable to the smallest things. Rather, we should have anticipated that many of these concepts would become less and less appropriate as the size of the objects under study diminished drastically. By the time we reach the tiniest underlying lumps of stuff out of which everything else is constructed, there is no reason at all why concepts like position and velocity, which grew out of our experience with comparatively enormous things like planets, automobiles, baseballs, or mosquitoes, should continue to have any direct relevance.

The quantum theory maintains that although it is possible to do things we call "measuring the position" or "measuring the velocity" of an electron, it is, in fact, no longer correct to regard those acts as ascertaining the value of some property inherent in the electron ("its position" or "its velocity"). Instead, such acts of measurement are rather intricate processes, which, to be sure, involve the electron but also involve whatever apparatus we subject the electron to in our efforts to learn "the value of its position" (or velocity). To try to abstract from this procedure an abstract position (or velocity) associated only with the electron, rather than jointly with the electron *and* the method we have used to ascertain that position (or velocity), is simply wrong. In the traffic cop analogy, the corresponding point of view would be that the reported speed of the car was not a property of the car at all but a joint manifestation of the car, the police helicopter, and the radar gun.

This is the kind of thing Heisenberg had in mind when he denied the existence of "an objective real world whose smallest parts exist objectively ... independently of whether or not we observe them." Pascual Jordan, another of the founders of the quantum theory, makes the same point:

> Observations not only disturb what has to be measured, they produce it. ... We compel [the electron] to assume a definite position. ... We ourselves produce the results of measurement.[12]

This idea, though it may surprise you the first time you encounter it, is extremely simple and not at all mysterious. If the things we are observing are so minute and delicate, and if the processes available for us to observe them are unavoidably clumsy and disruptive, then there may indeed be no legitimate way to abstract from an act of observation something we can associate only with the particle we observe, independently of how we choose to observe it.

Einstein, however, didn't like this. He wanted properties to be out there whether we observed them or not. The moon should be there whether or not somebody looked. It should not require observation by a person or a mouse to bring it into existence. After wrestling with this problem for some time, he finally acknowledged that the uncertainty principle did indeed impose practical and unavoidable limitations on how much information could be collected about physical properties, but he refused to conclude from this that the properties were not "there" at all, only being created by the act of measurement.

The quantum theorists, on the other hand, maintained precisely this. One might wonder why they held so extreme a position. It does, after all, seem a big additional step, and quite an unnecessary one, to go from the discovery that one cannot measure with perfect precision both the position and velocity of an electron to the assertion that an electron cannot have both a position and velocity. But several facts supported this more extreme position. Most important (as this essay will demonstrate), if one tentatively adopted the position that an electron always did have a definite position and velocity, whether or not either was measured, then one could invent situations in which it was extremely difficult if not impossible to imagine just what those unmeasured positions and velocities could possibly be in order to account for what was actually found when certain other positions (or velocities) were measured. In view of such difficulties, it seemed simpler to take the view that position and velocity were concepts evolved to describe matter on an entirely different scale, and that far from it being surprising that they no longer could both together be applied to electrons, perhaps the more remarkable thing was that one or the other of them still could.

A more abstract reason for being willing to sacrifice the notion of properties residing in the object is that the mathematical formalism of the quantum theory, a subject we shall have mercifully little to say about here, describes electrons as objects so unlike the kinds of things we usually deal with that it simply makes no sense to view them, within the framework of that theory, as endowed with definite positions and velocities. Instead, they are things to which the concepts of position and velocity only apply in a vague and shadowy way that precisely embodies and accounts for the uncertainty relations of Heisenberg. If you believe there is nothing more to electrons than what the quantum theory has to say about them, then you have to believe it makes no sense at all to provide them with both position and velocity.

For Einstein, the answer to this was simple. If the quantum theory cannot describe both the position and the velocity of an electron, then the quantum theory does not provide a complete representation of what the electron actually is. Electrons are more than what the quantum theory discloses them to be. The world is richer than our current attempt to understand it, and eventually, difficult as it may now appear, we will come up with a description that is powerful enough to incorporate everything that is really there.

In support of this conviction, Einstein invented a very ingenious experiment, which forced the quantum theory into maintaining that a previously nonexisting property in one region of space (region B) was brought into existence by a measurement made in a second region of space (region A) far away from and entirely unconnected with region B. To insist under such circumstances that it was the measurement at the remote place A that brought into existence the property at B was manifestly absurd to Einstein. For in contrast to the many examples invented by Bohr and Heisenberg, there was, in this case, no possibility of the measurement process at A *mechanically* disrupting things in the faraway region B. To Einstein this made the inadequacy of the Bohr–Heisenberg position absolutely evident:

> That which really exists in B should ... not depend on what kind of measurement is carried out in part of space A. It should also be independent of whether or not any measurement at all is carried out in space A. If one adheres to this ... one can hardly consider the quantum theoretical description as a complete representation of the physically real. If one tries to do so ... one has to assume that the physically real in B suffers a sudden change as a result of a measurement in A. My instinct for physics bristles at this.[13]

Most reasonably, Einstein maintains that properties in region B cannot be affected by what is going on far away in region A. Since the quantum theory nevertheless refuses to assign them any reality until a measurement is made in region A, it is clear that there are real things (namely those properties in region B prior to the measurement in region A) that the quantum theory fails to tell us about. Therefore its description does not provide a complete picture of physical reality.

The only way to avoid this conclusion is to continue to insist that the properties in region B were brought into existence by the measurement in region A, even though region A is far from region B, utterly unconnected with it, and incapable of influencing what is going on there through any known mechanism. This alternative causes Einstein's instincts to "bristle," and indeed, elsewhere he writes:

> I cannot seriously believe ... [in the quantum theory] because the theory cannot be reconciled with the idea that physics should represent a reality in time and space, free from spooky actions at a distance.[14]

It is these "spooky actions at a distance" ("*spukhafte Fernwirkungen*") that Einstein found most unacceptable in the quantum theory. By denying that the properties in region B were there until the measurement had been made in faraway region A, the quantum theory seemed to be embracing unacceptable remote connections. Einstein would have none of it. The properties must have been there all along.

Most physicists had little patience for such scruples. After all, the kinds of properties Einstein wanted to exist were properties such as the simultaneous position and velocity of a given electron, which he himself was willing to acknowledge could never, as a matter of principle, both be measured together. Nevertheless, he insisted, they must both have values. Wolfgang Pauli, another of the great early practitioners of the quantum theory, expressed himself pointedly on this matter:

> One should no more rack one's brain about the problem of whether something one cannot know anything about exists all the same, than about the ancient question of how many angels are able to sit on the point of a needle. But it seems to me that Einstein's questions are ultimately always of this kind.[15]

The experiment with which Einstein launched this attack on the quantum theory was presented as a *gedanken* experiment ("thought experiment") – a hypothetical sequence of events about which the quantum theory makes quite definite predictions. To make his point it was quite unnecessary actually to perform the experiment (which would also have been quite difficult to do in practice) since the purpose of the exercise was to challenge the quantum theory on the basis of what it *predicted* the outcome of such an experiment would be. Nobody, Einstein or anybody else, thought that if the experiment were done the outcome would disagree with the predictions of quantum theory. If it had disagreed, of course, the quantum theory would have proved to be wrong, and the question of whether or not it was complete, rendered moot.

The *gedanken* experiment was published in a 1935 article with

Boris Podolsky and Nathan Rosen[16] and was known for some time as the "Einstein–Podolsky–Rosen paradox," an interesting misnomer, since the paper claims to be nothing more than an argument that the quantum theory gives an incomplete description of physical reality. The widespread use of the term *paradox* indicates that a mere ten years after its full formulation, physicists were already so convinced that the quantum theory did give a complete description of the physically real that an argument to the contrary could widely be perceived as paradoxical.

More recently, people speak instead of the "Einstein–Podolsky–Rosen experiment." This dropping of the "*gedanken*" reflects several developments since 1935. In the earlier 1950s David Bohm produced a new version of the Einstein–Podolsky–Rosen argument. In the original *gedanken* experiment, by choosing to measure either the position or the velocity of a particle in region A, one could learn either the position or the velocity of a second particle in region B. Since the measurements in A could not disturb or disrupt the particle in region B, Einstein, Podolsky, and Rosen concluded that the particle in region B must have had both its position and velocity all along, thereby challenging the most famous example of the Heisenberg uncertainty relations, which explicitly prohibited, and indeed insisted it was meaningless to contemplate, the assignment to a single particle of both a position and a velocity. Bohm's variant resembled the original in that one learned the value of one of several mutually unknowable properties of one particle in region B by measuring an appropriately related property of a second particle in region A. However, the configuration in which these properties would have the necessary correlations to apply the Einstein–Podolsky–Rosen argument was much easier to arrange than in the original Einstein–Podolsky–Rosen paper; the measurements of the properties were the kinds of measurements that people had become very good at making in other contexts for quite some time; and Bohm's example was free of certain distracting but ultimately irrelevant complications in the original

Einstein–Podolsky–Rosen argument, associated with the fact that the properties they applied their argument to were, specifically, position and velocity.

Bohm's version of the argument was not only more clear-cut than the original; it allowed for the measurement of a considerably wider choice of properties than the two (position or velocity) considered by Einstein, Podolsky, and Rosen, though Bohm at the time did not take advantage of this added flexibility. More than a decade later, John Bell pointed out that the additional flexibility had some very dramatic consequences for the entire discussion. In an analysis[17] that has since gone under the name of "Bell's theorem" he showed that although some of the quantum theoretic predictions for Bohm's version of the Einstein–Podolsky–Rosen experiment did indeed appear to support Einstein's argument that certain physical properties had to have definite values whether or not they were measured, there was nevertheless no conceivable way to assign values to those unmeasured properties that could account for some of the other quantum theoretic predictions for that experiment. Bell thereby demonstrated that, Pauli to the contrary notwithstanding, there were circumstances under which one *could* settle the question of whether "something one cannot know anything about exists all the same," and that if quantum mechanics was quantitatively correct in its predictions, the answer was, contrary to Einstein's conviction, that it does not.

The importance of Bell's contribution to the dispute was profound. Prior to his analysis, the controversy was between Einstein and the metaphysical superstructure of the quantum theory. While allowing that quantum mechanics predicted correctly the results of experiments, Einstein maintained that certain properties had to have values whether or not they were measured, notwithstanding the fact that the quantum theory maintained that there was no meaning to such an assertion – that it was merely a "misuse of language." One could pick sides as one wished in such a dispute, confident that the issue was one of taste, judgment, or esthetics.

After Bell's analysis, the situation was transformed. If the quantum theoretic predictions for a conceptually very simple series of experiments were, in fact, correct, then Einstein's position had to be wrong, refuted not by debatable metaphysical principles promulgated by Bohr and Heisenberg but as a direct consequence of the data produced in those experiments. If the properties Einstein believed had values really did have them, then the description of physical reality provided by the quantum theory was not just "incomplete." It was wrong.

This was a sufficiently dramatic turn of events to inspire many people to attempt to carry out the Bohm–Einstein–Podolsky–Rosen *gedanken* experiment as an actual laboratory experiment, and a series of such efforts, culminating in some beautiful experiments in Paris by Alain Aspect and collaborators in the early 1980s,[18] unambiguously confirmed the quantum theoretic predictions, thereby putting an end to the Einstein–Podolsky–Rosen argument, not by decree of the founding fathers, but because it was inconsistent with the data.

Aside from making the Einstein–Podolsky–Rosen argument accessible to direct experimental test, another very attractive feature of the twist Bell gave to the argument is that by transforming it from a dispute about the proper interpretation of the quantum theory into a question that could be raised and settled in the laboratory, he made it possible to describe the controversy in a manner accessible to people with no knowledge whatever of the quantum theory and, for that matter, no training in physical science. The remainder of this essay is devoted to such an exposition.

I shall describe a very simple form of Bohm's version of the Einstein–Podolsky–Rosen experiment that contains within it both the very compelling argument of Einstein, Podolsky, and Rosen that certain properties have to exist and the subsequent demonstration by Bell that they cannot. The import of the experiment can be grasped by following an argument no more intricate than one might find in the puzzles department of a Sunday newspaper. The point is not the technical and rather erudite fact that the experiment demonstrates the impossibility of making the quantum theory the more complete representation of nature that Einstein wished it to be. The point that should make the experiment of interest to those who are neither physicists nor scholars is that the world behaves in a manner that is exceedingly strange, deeply mysterious, and profoundly puzzling.

The experiment, reduced to its bare essentials, will occupy us in Section II, where I shall present it as a collection of "black boxes" that behave in a simple but utterly baffling manner. In Section III, I shall give some indication of what actually goes on inside the black boxes, and how this behavior is related to the argument of Einstein, Podolsky, and Rosen. In Section IV, I shall describe some of the attitudes one can adopt in the face of these strange goings-on.

II. A remarkable device

We shall examine a very simple device, which behaves in a manner that appears at first glance neither remarkable nor even interesting. After describing the device and how it operates, I hope to convince you that its superficially bland behavior is, in fact, so mysterious that you may well conclude that I have been lying, and that such a contraption is impossible. This puts me, as an expositor, in a difficult position. If I succeed in explaining how the device works and why it is strange, then you may well decide not to believe me. Conversely, if, after reading this section, you are not suspicious of what I have told you, then chances are you have not followed my argument. ("If you do not get *schwindlig* ... then you have not really understood it.") I therefore begin by elaborating on the proper attitude to take toward surprising claims.

Suppose you had grown up on a desert island, ignorant of contemporary technology, and one day I washed ashore from a shipwreck, maintaining that I came from a land where there are small boxes you can carry around with you that can imitate human speech, sing songs, or produce the sound of many different musical instruments all playing together. You would not believe me, and this would be a pity, for you would thereby be deprived of one of the great pleasures of life: the contemplation of astonishing facts. The validity of this activity, of course, depends critically on whether what is being contemplated is indeed true. Had I asserted instead that there were boxes you could attach to people's foreheads which would then announce their name and date of birth, this would be at least as unbelievable, and equally worthy of contemplation, were it not entirely false.

Clearly it would be better to have something more than my mere assertion that what I am about to tell you is true. In the matter of the tape player, since I know a fair amount of physics, I could probably convince you that such a thing was possible by describing how it worked. I could not do this with the mind-reading machine. To be really persuasive about the tape player, however, I might have to take years of your time, requiring you to master various unfamiliar branches of science and mathematics, work through homework problems, and generally abandon everything else you might earlier have found to distract you on the island. In the end you might be persuaded, and you could then return to contemplating the wonders of a box that sang, confident that I was telling you the truth.

It would be far simpler had I managed to rescue a tape player before my ship went down. I could then have handed it to you, you would have pressed the button, and the box would have sung. You would not need to trust me or learn electronics. You would be convinced that a box can sing, because the box did sing.

What I am now about to describe is also interesting only because it seems unbelievable, but for that very reason it is only interesting if you believe it. The obvious way to convince you – to accompany every copy of this book with a working version of the device I shall describe – is impractical. It would be impractical even if device technology were as far advanced as tape player technology, but in fact the device is far more difficult to construct, and only recently have people been able to make versions that begin to approach the ideal behavior we shall examine below. Indeed, if I conducted you to the laboratory in Paris where the most impressive version of the device to date has been assembled and operated, you could still insist, were you very stubborn, that the astonishing behavior has not yet been irrefutably demonstrated, and I could not convince you otherwise.

That being so, why do I believe that I am telling the truth when I maintain that in principle (and, given some tiny fraction of the budget of the Strategic Defense Initiative, in practice too) a device that behaved like the one I will describe could indeed be constructed? My confidence is not primarily based on the fact that people have been improving on the earliest attempts and getting closer and closer to ideal behavior (though that is certainly worth noting). The real basis for my conviction is that the device exploits one of the most elementary of the enormous array of rich and intricate phenomena that the quantum theory has revealed over the past sixty years. If the device did not behave as I shall describe, the most successful scientific theory we have ever known would crumble, and our knowledge of the world would lie in ruins. The more than sixty years of phenomenal success the theory has enjoyed, on which most of modern technology rests, would have to be regarded as an accident – a lucky, but fundamentally unsound guess.

If you nevertheless persist in skepticism, then you should still find what follows interesting as an example of the extraordinary state modern physics has got into, that sane, reputable, and extremely distinguished people can publicly go around saying the kinds of things I am about to say. But they do. Indeed, most physicists have so absorbed the lessons of the quantum theory that they do not even regard any of the remarkable things I shall say below as unusual or disturbing.

It would be unfair, however, to offer you only my bald assertion of how the device behaves, so in Section III I shall attempt to describe some of the ingredients that go into its construction. Meanwhile, let us adopt the fiction that the device really is here, sitting in front of us, like the tape player I might have been lucky enough to bring with me to the desert island. Let me call your attention to its important features and the manner in which it operates.

The device: Three unconnected parts

After those prefatory warnings, you will be disappointed to see that the miraculous device is as depicted in Figure 1(a). It consists of three parts. Two of them (labeled "A" and "B") are for all essential purposes identical and are going to function as detectors, in a manner to be made precise below. The labels are intended to evoke Einstein's two faraway regions A and B, and the distance between them is indeed vast on the atomic scale (many meters in some of the actual experiments). The third part (labeled "C") sits between the two detectors A and B and is responsible for sending out something (represented very schematically by a blob accompanied by an arrow indicating its direction of motion) that the two detectors subsequently detect. Since whatever it is that sets off the detectors originates at C, we shall refer to C as the "source."

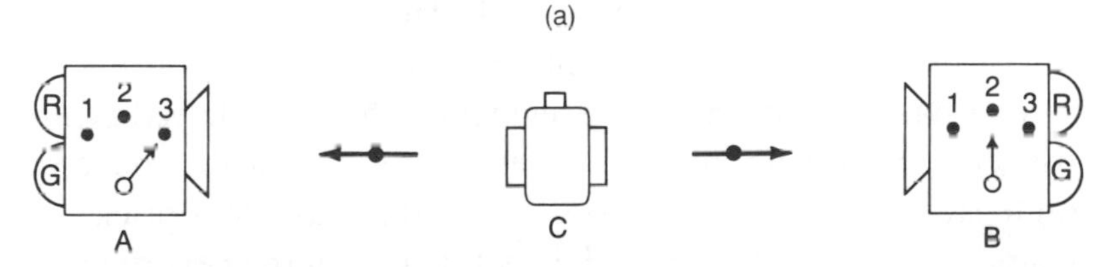

Fig. 1. A schematic representation of the astonishing device. The two detectors A and B are at the left and right of the pictures. The switches are set to 3 on A and 2 on B. The bulbs on the far sides of the detectors are colored red (R) and green (G). On the near sides of the detectors are two horns. Whatever comes from the source (C) passes into the horns to trigger the detectors. The source is in the middle with its button on top. Whatever travels to the detectors after the button is pushed emerges from the two portals on the sides and is represented in Fig. 1(a) by black dots with arrows pointing toward their destinations. Fig. 1(b) shows things just as a run of the experiment is finishing in which A ended up set to 3, B to 2, A flashed red, and B flashed green. The results of this run are entered into the record as "32RG."

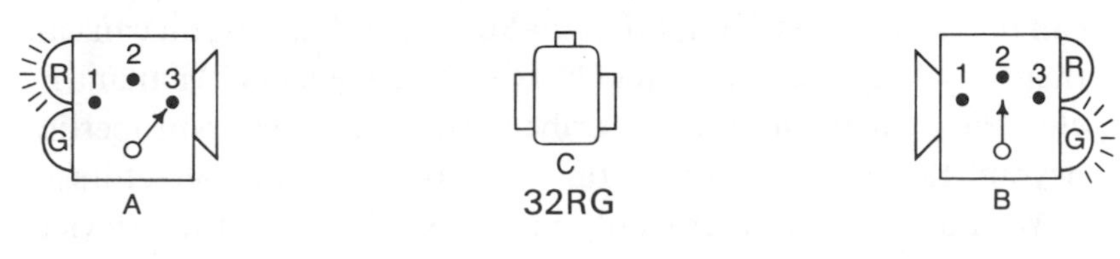

Each detector has a switch that can be set to one of three positions, labeled "1," "2," and "3." Each is also equipped with two light bulbs, one red and one green. When a detector detects something either its red light or its green light flashes. Each detector does this no matter how its switch is set, though whether it flashes red or green may well depend on the switch setting. The source C has a button on top. To operate the device, after setting the switches on the detectors – we'll come back to how to set them – you simply press the button. Shortly thereafter one or the other of the lights on each detector flashes. We call this a single "run" of the experiment.

That's all there is to it: first, set the switch on each detector; second, push the button on the source; third, note which color flashes at each detector. The full experiment consists of many such runs, for various settings of the switches. The information we record from each run is how the switch on each detector was set, and what color light flashed on each detector.

The detectors only flash their lights after the button on the source has been pushed. The evidence that the detectors are triggered by something that comes to them from the source is quite straightforward. If an obstacle – for example a brick – is placed between the source and detector A, then the next time the button is pushed and on all subsequent runs detector A no longer flashes, though detector B continues to flash as before. If the brick is removed, both detectors flash again. If the brick is between the source and detector B, then it is B that refuses to flash after the button is pushed. Evidently, then, pushing the button permits things to emerge from each side of the source C, which travel to detectors A and B and result in the flashing of the lights. Just what those things might be is unimportant for grasping the mysterious character of the device. Simply to have a name for them, we shall call them "particles," but if that makes you unhappy, you can call them anything else you prefer ("waves," "messengers," "bananas," etc.). We shall do our best to ascribe to these particles no properties beyond those we can infer from the flashing of the detectors.

We now come to a critically important feature of the device.

Aside from the fact that something evidently travels from C to A and something else from C to B each time the button is pushed, there are no connections between the source and either detector, and no connections between the two detectors. By this I simply mean that there are no mechanical connections such as strings, wires, rods, or pipes, no electromagnetic connections such as telephones, radios, radar, or light signals, no acoustic connections such as horns, bells, clicks, or whistles, and, in short, no relevant connections of any kind whatsoever.

"Ah," but you may say, "couldn't one detector signal to the other through the flashing light?" No, that can be blocked. The lights are at the far end of the detector, and one can put up barriers so that no light leaks out to the other detector. If you are still suspicious, we can replace the lights by marks ("R" for red and "G" for green) on a carefully shielded tape at each detector that records the results of successive runs. A more serious worry is irrelevant connections, which are hard to avoid. All three parts might be sitting on top of the same table, for example. But there is nothing in the design of the parts that could take advantage of such a connection to signal from one to another by inducing and detecting vibrations in the tabletop.

From my insisting so on the absence of connections, you may conclude that the wonders to be revealed by the operation of the device can only be appreciated by experts on connections or their absence. This is not the right point of view, for two reasons. First of all, if the crash program to build the device were successful and we were standing before it, you could pick up the parts, open them up, and poke around as much as you wanted. You would find no connections. Neither would an expert on hidden bugs, physicists called in as consultants, or professional debunkers of fraudulent claims. Second, when I describe in Section III the actual physical mechanism that makes the device work, whether or not you follow the discussion in detail, I would urge you to note that at least one thing is entirely evident from that description: the operation of the device does not rely in any way on connections between its parts.

But could there be unknown connections? How does one know that the parts are not connected by transmission of hitherto unknown "Q-rays" and their subsequent detection by currently unrecognizable "Q-detectors"? One can only offer affidavits from the manufacturer testifying to an ignorance of such "Q-technology" and, in any event, no such intent. Nevertheless, if you are stubborn about it, there is no way to rule out conclusively the possibility of connections – "spooky actions at a distance." And in fact the proper point of view to take is that the wonder and glory of the device are precisely that its operation impels one to doubt all the assurances from one's own eyes and hands, technical experts, and professional magicians that there are no connections between its parts. I can only insist that there are no connections that suspicious lay people or experts of broad erudition and unimpeachable integrity can discern. If you still find yourself questioning this after thinking about the data the device produces, then you will have grasped one of the deep mysteries of the atomic world.

The data produced by the device

At the beginning of each run the switches at each detector are set randomly and independently to one of the three positions, #1, #2, or #3. This could be done, for example, by rolling one member of a pair of dice at each detector, setting that detector to #1 if its die comes up 1 or 2, to #2 if the die comes up 3 or 4, and to #3 if the die comes up 5 or 6. Any other method will do just as well, provided that it leads, first, to each of the three settings of each detector being equally likely and, second, to no relation between the settings at the two detectors. Thus, one could use random numbers generated by two computers, one at each detector, to determine the settings; formulae based on stock market quotations at one detector, and baseball scores at the other; or any number of other methods.

We characterize each run of the experiment by the pair of numbers that specify the settings of the two switches. Thus a "12" run is one in which the switch on detector A ends up at position 1, and that on detector B, at position 2; in a "31" run, A is set to 3 and B to 1; in a "33" run both switches end up set to position 3. Altogether, there are exactly nine different possible types,

11 12 13 21 22 23 31 32 33,

each of which occurs about as often as the others over a long series of runs.

Having thus set the switches at the detectors in the most unimpeachably random manner we can devise, we push the button at the source and note the color of the light that subsequently flashes at each detector. A run in which both lights flash red is recorded as an "R" run, "G" indicates both lights green, "RG" means red at A and green at B, and "GR," green at A and red at B.

The information recorded after each run consists of the switch settings and the colors of the lights that flashed. Thus we would write "32RG" in our data book after a run in which the switch at detector A was set to 3, B was set to 2, A flashed red, and B flashed green (Fig. 1(b)); "11RR" would mean both switches were at 1 and both lights flashed red; "13GR" would mean A's switch was at 1, B's at 3, A flashed green and B, red.

Having thus recorded the result of one run, we go on to the next, randomly resetting the switches, pushing the button, and writing down the new switch settings and the new colors that flash at A and B. We do this over and over again, filling up many volumes with our two-number–two-letter symbols, until we have accumulated enough data to satisfy the most demanding statisticians. Let us avoid the technical question of how much data this might be. The point is simply that because the data has a statistical character, one could always blame the outcome on a series of accidents or coincidences. The more runs one has, however, the more implausible such an explanation becomes. By the time the number of runs has risen into the millions, the series of coincidences required to account for the outcome begins to verge on the miraculous. When this miracle becomes even more astonishing than the amazing behavior we are examining, we can abandon the explanation that appeals to persistent and systematic coincidences.

When we have completed some enormous number of runs, we can put the device aside and turn to studying the accumulated data. A typical page of such data (representing a tiny fragment of the total) is shown in Fig. 2(a); it gives the settings of the switches and the colors that might have flashed in about 100 runs. The entire body of data (from our millions and millions of runs) has two (and only two) very important features, which are easy to illustrate using the fragment of data displayed in Fig. 2(a):

First feature of the data. Consider only those runs in which the switches on both detectors happen to have ended up with the same setting. Since there are nine equally likely pairs of settings (11, 12, 13, 21, 22, 23, 31, 32, 33), but only three (11, 22, and 33) with both detectors set the same, we are considering here about a third of all the runs. Fig. 2(b) contains precisely the same fragment of

Fig. 2(a). A typical page from the voluminous record of many (perhaps millions) of runs of the device. Each set of two numbers and two letters represents the outcome of a single run. Reading such a set from left to right, the two numbers give the switch settings at A and B, and the two letters give the colors that flashed at A and B.

11RR	**33RR**	**11RR**	**12GR**	**12GR**	**32GR**
32GR	**11GG**	**23RG**	**32RR**	**22GG**	**11RR**
12RG	**23GG**	**31GR**	**13GR**	**11RR**	**11GG**
23RG	**21RR**	**11GG**	**32GR**	**31GR**	**22RR**
31GR	**11GG**	**23GR**	**12GR**	**31RG**	**12GR**
11GG	**23RG**	**33RR**	**33GG**	**31RG**	**13GG**
21GR	**21RG**	**31GR**	**13RR**	**13RR**	**31GR**
23RR	**13GG**	**11RR**	**31RR**	**11GG**	**13RG**
11GG	**21GG**	**12GR**	**21RG**	**11GG**	**32GR**
31GG	**12GR**	**32GR**	**31RR**	**33RR**	**32RG**
12RG	**22GG**	**13RR**	**21GR**	**11GG**	**23GR**
23RR	**13RG**	**11GG**	**21RR**	**11RR**	**23RG**
12RG	**31RR**	**23RG**	**22GG**	**23RG**	**31RR**
31GR	**21RG**	**13RR**	**31RG**	**31GR**	**32GR**
13RR	**22GG**	**31GG**	**23GR**	**12GG**	**22RR**
22RR	**23GG**	**32RG**	**21RR**	**23RG**	**12RG**

the data as Fig. 2(a) but displayed to emphasize the 11, 22, and 33 runs. Notice that in every such run, the lights have flashed the same color. What that color was can vary from one such run to the next, and in fact over many such runs RR and GG appear equally often, but when the switch settings are the same one *never* encounters RG or GR. This feature characterizes not only the sample data of Fig. 2(a) but all the data collected in our millions of runs:

1. When the switches have the same setting, the lights always flash the same color.

Second feature of the data. Figure 2(c) again displays the same data as Fig. 2(a) but now displayed to emphasize the second feature of the data, which emerges when one examines all of the runs, without regard to how the switches were set. If you examine only the pairs of colors that flashed – RR, RG, GR, or GG –

Fig. 2(b). The same typical page of data as displayed in Fig. 2(a), but displayed here to emphasize those runs (about a third of them) in which the switches ended up with the same setting (11, 22, or 33 runs). In these special runs the only color combinations that are ever observed are RR and GG. One never sees RG or GR. This illustrates the first important feature of the data: The lights always flash the same color when the switches have the same setting.

11RR	**33RR**	**11RR**	12GR	12GR	32GR
32GR	**11GG**	23RG	32RR	**22GG**	**11RR**
12RG	23GG	31GR	13GR	**11RR**	**11GG**
23RG	21RR	**11GG**	32GR	31GR	**22RR**
31GR	**11GG**	23GR	12GR	31RG	12GR
11GG	23RG	**33RR**	**33GG**	31RG	13GG
21GR	21RG	31GR	13RR	13RR	31GR
23RR	13GG	**11RR**	31RR	**11GG**	13RG
11GG	21GG	12GR	21RG	**11GG**	32GR
31GG	12GR	32GR	31RR	**33RR**	32RG
12RG	**22GG**	13RR	21GR	**11GG**	23GR
23RR	13RG	**11GG**	21RR	**11RR**	23RG
12RG	31RR	23RG	**22GG**	23RG	31RR
31GR	21RG	13RR	31RG	31GR	32GR
13RR	**22GG**	31GG	23GR	12GG	**22RR**
22RR	23GG	32RG	21RR	23RG	12RG

without paying any attention to the pairs of numbers giving the switch settings, you will find no regular pattern at all. The sequence is entirely random, each of the four possibilities occurring about a quarter of the time. In particular, the lights flash the same color (RR or GG) as often as they flash different colors (RG or GR). This feature of the data is also borne out in all our millions of runs:

> **2. In all runs together (regardless of how the switches were set), the same colors flash as often as different colors.**

This is all you need to know about the data. If you consider only those runs in which the switches ended up with the same setting (Fig. 2(b)), then you find a completely random list of the two symbols RR and GG, such as might have been produced if each entry were determined by tossing a fair coin and writing down RR for heads and GG for tails. But if you examine all the pairs of

Fig. 2(c). The same typical page of data as displayed in Fig. 2(a), now displayed to emphasize the colors and de-emphasize the switch settings. This illustrates the second important feature of the data: If the colors that flash in each run are examined without any regard to how the switches were set, then the pattern is a completely random list of the four symbols RR, RG, GR, and GG; in particular, the same colors (RR or GG) occur just as often as different colors (RG or GR).

11**RR**	33**RR**	11**RR**	12**GR**	12**GR**	32**GR**
32**GR**	11**GG**	23**RG**	32**RR**	22**GG**	11**RR**
12**RG**	23**GG**	31**GR**	13**GR**	11**RR**	11**GG**
23**RG**	21**RR**	11**GG**	32**GR**	31**GR**	22**RR**
31**GR**	11**GG**	23**GR**	12**GR**	31**RG**	12**GR**
11**GG**	23**RG**	33**RR**	33**GG**	31**RG**	13**GG**
21**GR**	21**RG**	31**GR**	13**RR**	13**RR**	31**GR**
23**RR**	13**GG**	11**RR**	31**RR**	11**GG**	13**RG**
11**GG**	21**GG**	12**GR**	21**RG**	11**GG**	32**GR**
31**GG**	12**GR**	32**GR**	31**RR**	33**RR**	32**RG**
12**RG**	22**GG**	13**RR**	21**GR**	11**GG**	23**GR**
23**RR**	13**RG**	11**GG**	21**RR**	11**RR**	23**RG**
12**RG**	31**RR**	23**RG**	22**GG**	23**RG**	31**RR**
31**GR**	21**RG**	13**RR**	31**RG**	31**GR**	32**GR**
13**RR**	22**GG**	31**GG**	23**GR**	12**GG**	22**RR**
22**RR**	23**GG**	32**RG**	21**RR**	23**RG**	12**RG**

colors produced in all the runs (Fig. 2(c)), you see a completely random list of the symbols RR, RG, GR, GG, such as might have been produced if each entry were determined by drawing a card from a well-shuffled deck and writing down RR, RG, GR, or GG depending on whether the card was a spade, heart, diamond, or club.

Analysis of the data

After all my talk about unbelievable behavior, you will be disappointed to learn that I have now told you everything. *Random colors in all runs; only the same colors when the switches are the same.* Is this profoundly mysterious? Is this the kind of thing that "nobody really understands"? Are you filled with disbelief and ready to accuse me of lying when I say there were no connections between the parts of the device? I suspect not. But you should be. The data I have described may well be the most profoundly disturbing ever to have been collected in an experiment, or derived from a valid scientific theory.

The first step toward comprehending the strangeness of the data is to focus entirely on its first feature. How can the lights always flash the same color when the switches have the same setting? Were there connections between the parts of the device, there would be any number of ways to arrange this. For example:

(*a*) If both detectors were connected to the source, then in the interval between setting the switches and pushing the button, each detector could send a message to the source, informing it of the setting of its switch. Suppose that when the button was pushed at the source it shot tiny red or green particles at the detectors, and that the color flashed by a detector was simply the color of the particle that triggered it. Then, in those runs in which the source learned that the switches had the same setting, it could take care only to shoot out particles of the same color, thereby ensuring that both lights flashed the same color. What could be more simple?

But the detectors are not connected to the source. Indeed, the concern that the device might exploit such hidden connections can be simply disposed of by arranging for the switches not to be set until after the particles have left the source, but before they have arrived at the detectors. If the experiment is thus refined, the character of the data does not change at all.

(*b*) If both detectors were wired together (even though neither was connected to the source) then they could communicate directly in each run of the experiment, agreeing on what lights to flash when the particles arrived. The only role of the particles in this case would be to signal the detectors, by their arrival, that the time had come to flash. The actual colors flashed could be entirely determined by the direct communication between the detectors. Since each detector would know in every run the setting of the switch on the other, they would know in which runs they had to flash the same color and could consult together in such cases to decide whether that color should be red or green.

But the detectors are not connected to each other, and therefore they lack the information needed to cooperate in so simple a way. The suspicion that they might be connected together in some obscure manner can be dealt with by arranging for one detector to flash well before the other (by moving one detector very close to the source, for example, and taking the other one very far away) and then not setting the switch on the far detector until *after* the light has flashed on the near one. This refinement does not alter the character of the data any more than in the previous case.

A stratagem for thwarting both concerns would be repeatedly and randomly to keep changing the switch settings at each detector at a rate so fast that no possible signal could get from one detector to the other or from either to the source before both switch settings had received several additional random flips. This tactic relies on the well-established fact that no signal of any kind can propagate faster than the speed of light, and the Paris version of the device actually incorporates such extremely rapid switch flipping, in an attempt to strengthen the (already strong) conviction that there are no connections between the three parts of the device.

Connections between the parts of the device would not only

make it easy to flash the same color when the switches had the same setting; with connections it is simple to make a device that produces *all* the data I have described, and several people have done just that, making a dishonest version of the device that works by cheating to demonstrate its behavior more vividly than any merely verbal description can. But in the real device there are no connections beyond whatever it is that moves from the source to the detectors after the button has been pushed. You have, I hope, agreed at least provisionally to accept my insistence on this point.

In fact, this one connection is quite enough to provide an entirely simple explanation for why the lights flash the same color when the switches have the same setting. Both detectors are triggered by particles that have originated in the same place (at the source C). Suppose each particle carries the following message to its detector: "Flash G if your switch is set to 1, G if set to 2, and R if set to 3," which we compactly summarize as "GGR." If each detector does as it is told upon receiving its message, then if both switches happen to be set to 1 both lights will flash green (giving the result 11GG), if both are set to 2 both lights will flash green (22GG), and if both are set to 3, both lights will flash red (33RR). If the particles always carry identical messages (whose contents can vary from one run to the next), then evidently the lights will always flash the same color when the switches have the same setting.

This explanation relies entirely on the one way in which the behavior of the detectors can legitimately be coordinated: each detector has available to it before it flashes something – a "particle" – that could easily be carrying flashing instructions from the source C. It is essential that the instruction sets carried by the particles specify the color for each of the three possible settings of the switches, since the detectors are not connected to the source, and there is therefore no way to know which, if any, of the three identical switch settings (11, 22, or 33) the particles might encounter at the detectors. Since any one of the three is always a possibility, and since the lights *always* flash the same color when the switches have the same setting, the particles have to provide instructions (each particle providing the *same* instructions) covering each of the three cases.

It is also essential that the particles carry such identical instruction sets in every run of the experiment. This is because the particles have no way of knowing how the switches are going to be set at the detectors when they arrive there. Although the switches have the same setting in only about a third of the runs, any given run might be one with equal settings. (Recall that the switches are set in a completely random manner and can, if you like, be set after the particles have left the source.) To maintain that perfect record – the lights *always* flashing the same color *whenever* the switches have the same setting – the particles have to carry instruction sets in every run of the experiment.

These considerations lay out the bare bones of what is needed for the lights always to flash the same color when the switches have the same setting, given that coordination between the detectors can only be provided by their being triggered by particles that have come from a single source. One can imagine any number of ways in which the information on how to flash could be carried. Most crudely, each particle could be a little box, containing within it a piece of paper on which is written one of the instruction sets,

RRR RRG RGR RGG GRR GRG GGR GGG,

these eight giving all the different possibilities. More sophisticated arrangements for conveying this information can easily be invented, without altering the underlying logical structure. The information might be carried by the shape of the particles: detectors could respond to little spheres by always flashing red (RRR); little cubes could cause detectors to flash green for settings 1 and 2 and red for setting 3 (GGR); detectors receiving cylinders could flash red for 1, green for 2, and red for 3 (RGR); and so on. The eight distinct instruction sets would simply be carried by eight distinct shapes of particle, and both particles in each run would always have the same shape.

You can invent your own coding schemes. The minimum any scheme must incorporate is that in one way or another each particle must carry enough information to determine how its detector is to flash for each setting of its switch, and that in each run both particles must carry exactly the same instructions. However that information is transmitted, it can be summarized abstractly as an instruction set such as RGR. We shall therefore speak of the instruction sets carried by the particles, with the understanding that this term need not mean anything as literal as a piece of paper upon which is written "RGR," nor anything as specific as a cylindrical shape which the detector is constructed to recognize as specifying RGR. The instruction sets are simply intended to represent the essential features of whatever (unknown) mechanism actually does permit the particles to carry the flashing information to the detectors.

At this point it would be well for you to pause and try to invent another mechanism for the lights to flash the same color when the switches have the same setting, a mechanism that respects the absence of connections but does *not* require the particles somehow to carry to the detectors information enough to determine how they flash for each setting of their switches. Unless you are well versed in the quantum theory, I can guarantee that you will fail (and even if you do know quantum mechanics, you will find it difficult unless you already know the elementary but subtle quantum mechanical phenomenon that underlies the operation of the device). The explanation in terms of instruction sets is so simple, contains so little unnecessary hypothetical structure, and seems so directly to be required by the behavior of the device, that it is difficult to see what could be wrong with it and (as you will discover if you try) virtually impossible to come up with anything else.

What is astonishing about the device and the data it produces is that this explanation in terms of instruction sets cannot be correct. Although it seems the only way to account for why the lights flash the same color when the switches have the same setting, it is utterly inconsistent with the second feature of the data

– the fact that in all runs of the experiment, regardless of how the switches end up, the same colors flash just half the time.

How can this apparently innocent behavior destroy the hypothesis of instruction sets? The first thing to keep in mind is that, as already noted, the particles must carry identical instruction sets in *every* run of the experiment – even those in which the switches end up with different settings – since the source cannot know how the switches are going to be set in any given run. Because any run *might* be one in which the switches end up with the same setting, the particles *must* carry identical instruction sets to ensure that the detectors *always* flash the same color whenever the switches have that setting.

We must therefore find a way to provide the particles with instruction sets that result in the same colors flashing in just half of all the runs of the experiment. At first glance this might not appear daunting. There are eight different instruction sets, which we are at liberty to assign to the particles in any way we wish. Surely it must be possible to find *some* way of shuffling them from one run to the next so as to lead to an entirely random pattern of colors flashing when all the runs together are examined without reference to how the switches were set.

Remarkably, there is not. The proof that there is not is known as "Bell's theorem." In the context of our device, the proof of Bell's theorem is so simple that the argument hardly seems to deserve so formal a title. It goes like this:

As a first step in trying to decide which instruction set the particles should be carrying in each run of the experiment, let us consider the patterns of flashing produced by each of the eight types of sets. We begin with those instruction sets in which both colors appear, of which we can take RGG as a typical example. Whenever both particles carry the instruction set RGG, the same color flashes if the switches are set to 11, 22, 33, 23, or 32. (The first case results in RR flashing; the other four result in GG.) The four remaining possible settings, 12, 13, 21, or 31, result in different colors (RG in the first two cases and GR in the last two). Thus when the instruction set is RGG, five out of the nine possible switch settings result in the same color flashing, and four of the nine result in different colors.

Since the switches are set randomly in each run of the experiment, each of these nine settings occurs equally often, so in all runs of the experiment in which the instruction set is RGG, the same color will flash 5/9 – about 55.5% – of the time. This means that the rule cannot be "always carry the instruction set RGG," for in *all* runs of the experiment the same color only flashes 50% of the time. So we must use some of the other instruction sets as well.

But now we are in serious trouble. Consider any of the remaining instruction sets in which both colors appear (GRG, GGR, RRG, RGR, GRR). In every one of these cases we can make the same simple enumeration of the possibilities as we did for the instruction set RGG. We will again conclude that the same color will flash for five of the nine switch settings and different colors for the other four. You can check this by going through each case, but that is hardly necessary, since all we used to reach our conclusion for the case RGG was that one color appeared once in the instruction set and the other color twice.

So *all* of the six instruction sets – GGR, GRG, RGG, RRG, RGR, GRR – result in the same color flashing a little more than 55.5% of the time, and there is thus no way, using these instruction sets alone, to explain the fact that the same color flashes only 50% of the time. But the only instruction sets left to us are RRR and GGG, and these result in the same color flashing *all* of the time. Using them only makes things worse!

We conclude that, regardless of how the instruction sets are chosen from one run to the next, if one examines the data from all runs of the experiment without regard to how the switches were set, one must find that the same color has flashed *at least* 55.5% of the time (and even more often than that, if the instruction sets RRR and GGG are used more than rarely). This conclusion is known as "Bell's inequality," and the argument we have just given that leads to it is Bell's theorem.

Bell's inequality tells us that regardless of how the instruction sets are distributed, in all runs of the experiment the same color must flash at least 55.5% of the time. But as the device actually operates, the same color flashes only 50% of the time. The device violates Bell's inequality. But Bell's inequality is a consequence of the existence of instruction sets. Therefore instruction sets cannot exist.

But instruction sets were the only explanation we were able to find for how the lights can flash the same color when the switches had the same setting. So how *can* they? Yet they do. That is the conundrum posed by the device – its sublime mystery.

We turn next to a glimpse of the actual mechanism that causes the lights to flash the same color when the switches have the same setting, while preserving the random character of all the data. You can decide for yourself whether that constitutes an adequate resolution of the mystery. One thing, however, should become absolutely clear: it is not an explanation that relies on connections, either between the detectors and the source, or between the two detectors. There *are* no connections.

III. Inside the device

My description of the actual physical processes underlying the device will necessarily be incomplete. It would be impractical to teach enough quantum mechanics to make the mechanism behind the device evident from a few basic principles, and you will have to accept my assurances throughout the discussion that certain things do indeed behave as I maintain. I shall develop the device in stages, starting with certain simple phenomena one can demonstrate with two pairs of polarizing sunglasses, available at any drugstore. Those phenomena raise questions, to which the next stage of the discussion provides rather plausible answers. The questions are about whether unobserved properties can have values and are thus of the kind described in Section I as quintessentially characteristic of the quantum theory. Our attempts to answer the questions will lead directly to a version of the Einstein–Podolsky–Rosen experiment which, when suitably packaged with switches and colored lights, produces precisely the behavior of the device. Whether or not you follow closely the line of argument leading to this culmination, please note when we get there that *there are no connections between the parts of the device*, which is without a doubt the most important point of the whole development.

We begin with some simple experiments with a beam of light and three identical transparent plastic disks of a rather special kind. Regard the beam of light as a shower of little particles – Einstein's photons. The plastic disks are made out of what is called polarizing material and can be acquired by purchasing two sets of polarizing sunglasses at the drugstore and removing the four lenses (you get a spare in this way). It is crucial that the sunglasses have polarizing lenses and not lenses of ordinary dark glass or plastic. A simple way to test for this will emerge shortly. The experiments reveal various properties of the disks and, more importantly, of the photons themselves.

Experiments with one disk

Nothing very interesting happens with a single disk. If we interrupt a beam of light with a disk, holding it perpendicular to the beam as in Fig. 3, some of the photons get through and some do not. Precise measurements reveal that the beam that gets through is half as bright as the original beam, indicating that half the photons are able to get through the disk. How do we know that the dimmer light doesn't mean that all the photons get through, but in some "weaker" form? One can use photon detectors that click when a photon enters them (unfortunately not available at the ordinary corner drugstore); when a disk is placed in front of such a detector, it clicks only half as often.

With the particular beam of light I have in mind (for example, the beam from an ordinary flashlight), this behavior is unaffected by rotating the disk in its plane about the direction of the beam (Fig. 3). This is no different from our experience with ordinary window glass and is mentioned here only because of the less familiar behavior when two disks are used.

Fig. 3. A beam of light from a flashlight, interrupted by a single polarizing disk. The intensity of the light that gets through the disk is cut in half, indicating that half the photons get through the disk and half are blocked. If the disk is rotated in its plane about the direction of the beam of light, this has no effect whatever on the intensity of the light that gets through. Half the photons get through and half are blocked, whatever the orientation of the disk.

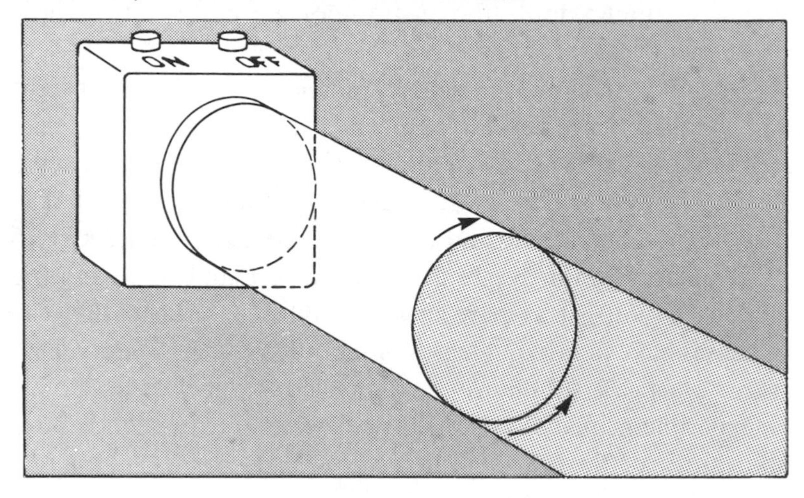

Fig. 4. A second disk is placed in the beam, identical to the first. How much of the beam does it let pass? A reasonable guess might be that it would again reduce the intensity of the beam by half, letting through only half the photons that get through the first disk. Instead, we find that the amount of light that gets through both disks depends on the orientation of the second disk with respect to the first. By rotating the second disk in its plane about the direction of the beam of light we change the amount of light emerging from the first disk that gets through the second.

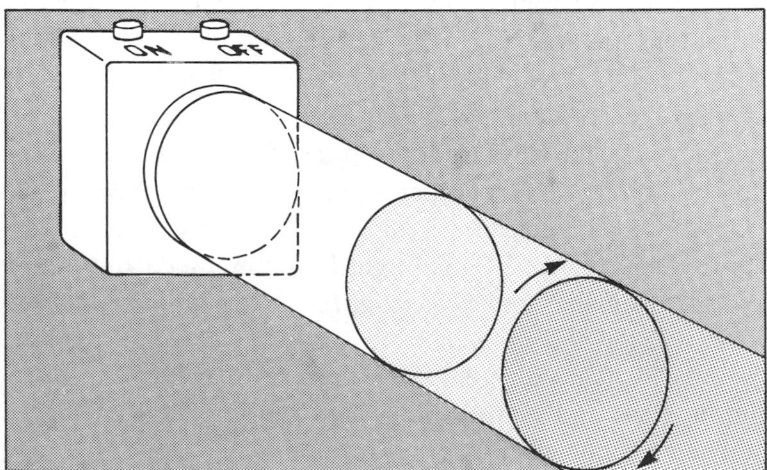

Experiments with two disks

Keeping the first disk in place, we interrupt the diminished beam of light coming through it with a second, also placed perpendicular to the beam (Fig. 4). Since the disks are identical, one might expect the second disk, like the first, to let through half the photons landing on it. This expectation is not fufilled. Since the disks are indeed identical – interchanging them makes no difference – the photons in the beam that got through the first disk must have a different character from those in the original beam. To explore this, let us examine in more detail what does happen when the second disk is set in place.

What happens depends on the orientation of the second disk with respect to the first. If, keeping the first disk fixed, we slowly rotate the second about the direction of the beam (Fig. 4), the intensity of the light getting through the second disk changes. There is an orientation of the second disk at which the intensity of the beam emerging from the first suffers no significant additional drop on passing through the second (Fig. 5). At that special orientation all the photons that passed through the first disk get through the second too. As we rotate the second disk past this special orientation, the intensity of the beam that gets through it drops. When we reach an angle 90° from the special orientation, the intensity has dropped to zero – no photons at all get through the second disk (Fig. 6). If we continue to rotate past 90° the intensity builds up again, until we reach 180°, at which point all the photons incident upon the second disk again get through it. Between 180° and 360° the pattern encountered between 0° and 180° repeats itself. (Quite generally, rotating a disk through 180° has no effect on its performance.)

I digress to remark that this is what to check when acquiring the two pairs of sunglasses. Place a lens of one pair directly in front of a lens of the other and look through the two. Then start turning one while keeping it in front of the other so your line of sight continues to pass through both. If the two-lens pair varies from transparent to almost entirely opaque as you turn one lens, then you have the right kind of sunglasses. If there is no variation in what comes through as you turn one lens, then at least one pair is not made of polarizing materials, and you should try others.

As what gets through the second disk depends on its orientation with respect to the first, it is important to have a way to specify this relative orientation. We draw a line on the first disk in some arbitrary direction. Fixing the orientation of the first disk, we now rotate the second until it is in the orientation that lets the most light through – i.e., the orientation of Fig. 5 that permits all the photons incident upon it from the first disk to pass. With the second disk in that special orientation, we draw a line on it parallel to the line on the first (Fig. 5). For future use we provide all our disks with lines after aligning them in this way with the first

Fig. 5. The second disk can be turned to an orientation at which all the light falling on it gets through: every photon that gets through the first disk also gets through the second. When the second disk is in that special orientation we draw parallel lines on both disks (in any orientation we like, as long as they are oriented the same way on both disks). Keeping the orientation of the first disk fixed, we repeat this procedure with all our other disks as second disk, turning each until it transmits all the light that gets through the first, and then furnishing it with a line parallel to that on the first. It is then found (unsurprisingly) that if the beam is interrupted by *any* two disks that have their lines parallel, then the second disk does not further diminish the intensity of the light that gets through the first.

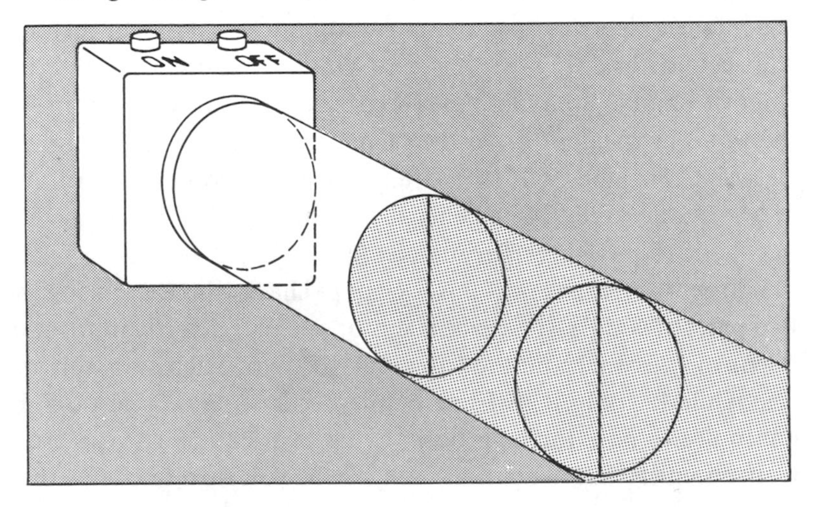

disk. We can then verify that if the line of any one disk is parallel to the line of any other then all the photons emerging from one get through the other.

Its line provides a way to specify the orientation of a disk in the plane perpendicular to the beam of light. We shall say that the orientation of a disk is the number of degrees its line is rotated away from the vertical. By the angle between two disks we shall mean the angle between their lines. (We always orient the *plane* of the disks perpendicular to the beam of light.) With this nomenclature we can state the results of the two-disk experiment as follows: if the angle between two disks is 0°, then every photon that gets through the first gets through the second as well, and the two disks together are as transparent as the first disk alone (Fig. 5). But if the angle is 90°, then no photon that gets through the first gets through the second, and the pair of disks together is opaque (Fig. 6).

Fig. 6. The second disk can be turned to another orientation (by turning it 90° beyond the orientation of Fig. 5) at which no light gets through it; every photon that gets through the first disk is blocked by the second. Because the second disk has been rotated 90° with respect to the first, their two lines are now perpendicular rather than parallel. Quite generally, when the lines on two disks are perpendicular every photon that gets through the first disk is blocked by the second.

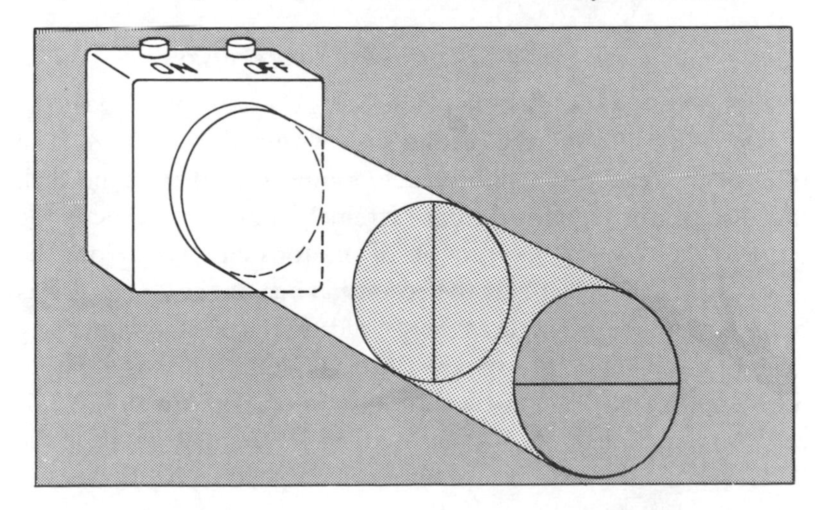

The behavior at intermediate angles is also easily characterized. When, for example, the disks are at 30°, three-quarters of the photons getting through the first disk also get through the second. When they are at 45°, just half get through the second, thereby, incidentally, behaving according to our original naive expectation but only for this special angle (Fig. 7). When the angle is 60°, only a quarter of the photons that get through the first disk also get through the second. The rule for a general angle α between the disks is that the fraction f of photons getting through the first disk that also get through the second is just the square of the cosine of α: $f = \cos^2 \alpha$.

If you have forgotten trigonometry, no matter! We shall be interested in particular cases rather than the general formula, and I have already given the fractions f for all the angles α that will be of interest in what follows:

Angle α	Fraction f
0°	100%
30°	75%
45°	50%
60°	25%
90°	0%

Experiments with three (or more) disks

Keeping the first two disks in place, we now interrupt the beam of light coming through the second disk with a third (Fig. 8). The rule for the fraction of photons emerging from the second disk that also get through the third is extremely simple: it is exactly the same as the rule for the fraction of photons emerging from the first disk that get through the second. Thus, if the third disk is oriented at the same angle as the second, all the photons emerging from the second get through the third. If the third disk is at 90° to the second, none of them get through; if the disks are at 45°, half get through; and so on. All that matters is the angle between the third disk and the second. The same is true of a fourth, fifth, or any

Fig. 7. The naive behavior guessed at in Fig. 4 can actually occur, but only when the angle between the two disks is 45°. For that special orientation, and only for that special orientation, half the photons that get through the first disk also get through the second.

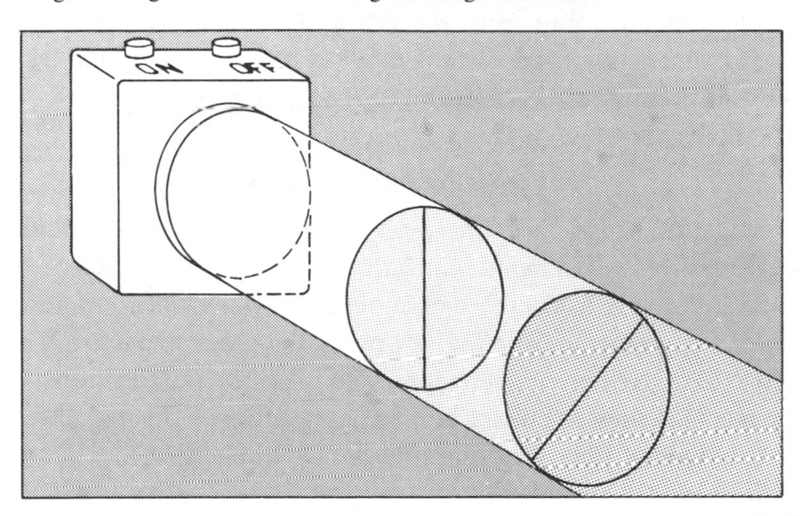

Fig. 8. A third disk is placed in the path of the beam. The fraction of photons emerging from the second disk that get through the third is entirely determined by the angle between those two disks, and is given by the same rule as gave the fraction of photons emerging from the first disk that got through the second. This simple rule has some surprising consequences.

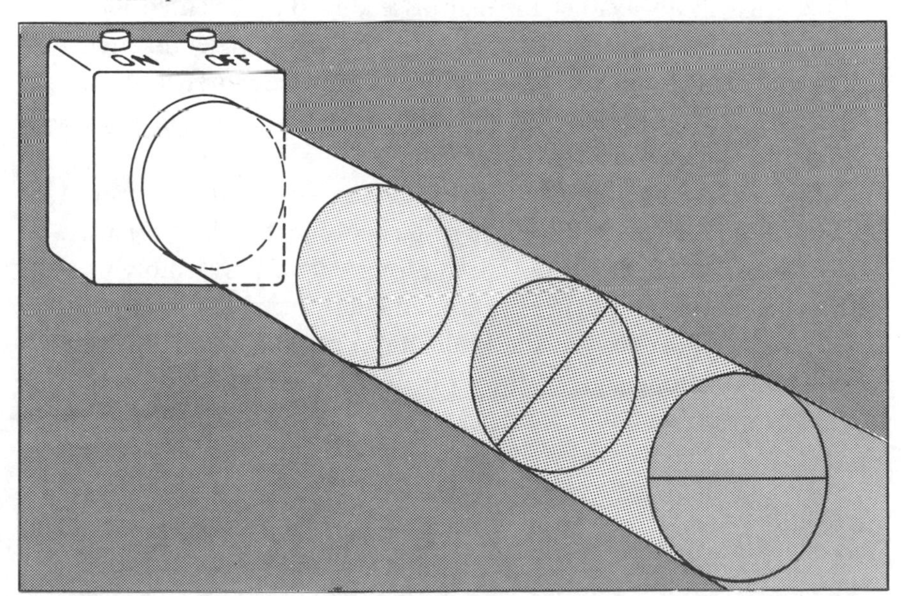

number of subsequent disks. Each new disk blocks a fraction of the photons emerging from the last disk already in place – a fraction that is entirely determined by the angle between that last disk and the new one, according to the same rule.

The polarization of photons

Already these simple facts can result in some very striking (though not astonishing) behavior, consideration of which I defer to introduce an important piece of additional nomenclature. Those photons that emerge from a disk oriented at a particular angle α shall be said to be "polarized along α" or to "have polarization α." The facts just described can be translated into statements about the behavior of polarized photons.

Consider a beam of photons interrupted by a disk oriented along α. By definition, the photons that get through the disk are polarized along α. If we now interrupt that beam with a second disk, also oriented along α, then since the angle between the two disks is $0°$, all the photons that got through the first will get through the second. Thus a photon that is polarized along a given direction will always get through a second disk oriented along that direction. Since none of a beam of photons that get through a disk will get through a second at a $90°$ angle to the first, a photon that is polarized in a given direction will never get through a second disk at $90°$ to that direction.

Under these two circumstances we can state with assurance how a photon will behave when it encounters a disk oriented along α: if the photon is polarized along α it will invariably pass through the disk; if polarized perpendicular to α it will invariably fail to pass through. We shall say that a photon "has a definite polarization along a given direction" if it is either polarized along that direction or polarized perpendicular to that direction.

Do photons have definite polarizations along all directions?

Is the fact of being polarized along a direction α a property impressed upon a photon by a disk oriented along α through which the photon has just passed, or was this polarization already possessed by the photon even before it entered the disk? In the first case the disk actively changes the nature of the photons it allows to pass; in the second case it is merely a passive filter, letting through photons that have the right polarization and blocking those that do not. (Note the similarity to the questions raised in Section I: Are the properties inherent in the object, or are they at least in part produced by the very act of testing for those properties?)

A photon that has just passed through a disk oriented along a given direction (which we take to be 0°) is by definition already polarized along 0°, and its subsequent passage through any number of disks oriented along 0° (or its failure to pass through a disk at 90°) is correctly viewed as merely confirming this polarization without altering it. But suppose the second disk is at 45°. Then half the photons passing the first disk will get through the second and half will not. Those that do get through the second will be polarized, again by definition, at 45°. If disks were merely passive filters, then of all the photons that got through the first disk half, in addition to being polarized along 0°, would also be polarized at 45° along the direction of the second disk, while the other half would also be polarized perpendicular to that second direction.

Of course we could have tested the photons emerging from the disk at 0° for polarizations in additional directions. Had we put in a second disk at an angle 30°, then (see the above table) 75% of the photons emerging from the first disk would have made it through the second. If the disks were passive filters, we could conclude that 75% of the 0° photons that passed the first disk were also polarized along 30°, while the other 25% were also polarized perpendicular to 30°. Since the second disk can be set at any angle, if the disks were passive filters each photon would have to possess a definite polarization along any direction at all.

It is easy to invent models of photons with this character. Figure 9 shows one such highly schematic view of a photon as it might appear coming directly out of the page toward a disk. It looks like a pie, divided up into two kinds of angular regions, indicated by black and white in the drawing. If the angle of the disk lies in the white region the photon gets through, but if it lies in the black region the disk blocks the photon. Different photons, in this view, could carry different patterns, but the crucial assumption is that every photon would carry some such pattern, thereby conveying to any disk set at any orientation enough information to determine whether or not the disk should let the photon through. Thus the photons that passed a 0° disk would all be white in the 0° direction. If those photons subsequently encountered a 30° disk, the ones that emerged would be those that were white in both the 0° and the 30° direction. If a third disk was

Fig. 9. A highly schematic model of the possible internal structure of a photon as it might appear to a disk the photon is approaching. (The circle here is the photon, enormously expanded in size for clarity; it is not a disk.) If a photon were decorated in this way the disk could decide whether or not to let it pass according to the following rule: If the line on the disk is along a direction pointing from the center of the photon into one of the white regions, then the photon gets through; but if the line on the disk is in a direction that points from the center into one of the black regions, then the photon is blocked. Note that various simple properties of photon polarization are satisfied by the coloring: Directions 180° apart all have the same color, as they must since the behavior of the disk is unaffected by a 180° rotation; directions 90° apart always have opposite colors, as they must since a photon polarized in a given direction invariably fails to get through a disk at 90° to that direction.

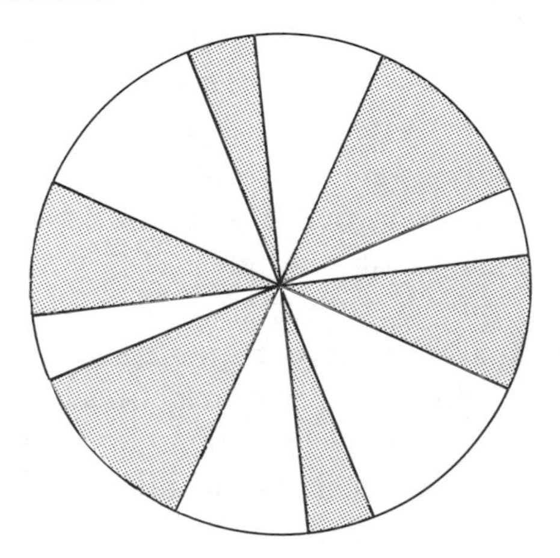

then encountered at 45°, the beam emerging from it would contain photons that were white in all three directions, 0°, 30°, and 40°. (Since photons polarized in a given direction are always blocked by a disk at 90° to that direction, it is essential that the pie have opposite colors at angles that are 90° apart, but that is easily arranged.)

The crucial thing in this or any other such model is that if the disks were merely passive filters, then each photon would in some way or other have to carry the information determining whether it would pass a disk at any direction whatever. (This is highly reminiscent of the "instruction sets" of Section II.) Precisely how photons would actually carry this information is something we would have to understand better if indeed the disks were passive filters.

It follows from the little I have already said, however, that the disks cannot be passive filters. Consider, for example, a sequence of three disks, at 0°, 45°, and 90° (Fig. 8). Half the photons emerging from the 0° disk get through the 45° disk, and half those emerging from the 45° disk get through the 90° disk, since all that matters is the angle (90° – 45° = 45°) between the last two disks, so a quarter of those getting through the 0° disk also get through the 90° disk. But we know that no photon polarized along 0° can get through a 90° disk. Therefore the 45° disk cannot simply have let through some of the 0° photons with their 0° polarizations unaltered. At least half of the photons emerging from the middle (45°) disk cannot be 0° photons. The (45°) disk is not only filtering photons. It is actively modifying at least some features of their polarizations.

This simple behavior is quite surprising the first time it is seen, and observing it is well worth the price of two pairs of polarizing sunglasses. (The "clip-on" kind are the cheapest and every bit as good, since all you need are the lenses.) One takes a sandwich of three disks at 0°, 45°, and 90°. The sandwich is transparent: 12.5% of the light incident upon one side of the sandwich gets through (each of the three disks cuts out half the light incident upon it and half of a half of a half is an eighth – 12.5%.) If, however, the central disk is removed, then the remaining two-disk sandwich becomes completely opaque (compare Figs. 8 and 10) – no light gets through at all. This is in startling contrast to our expectation that *more* light should get through, since we *removed* an obstruction.

The effect is particularly dramatic if, instead of shining light through the three-disk sandwich, you look through it. Since 12.5% of the light from the other side gets through, you can see quite clearly through the sandwich. After the middle disk is removed, however, nothing at all can be seen. By removing an obstruction you have made it harder to see through the three disks. Conversely, if the middle disk is reintroduced, still at its 45° angle, the sandwich again becomes transparent. By adding a disk to the pile you have succeeded in letting through more light.

The behavior is surprising because we seem unconsciously to expect the disks to be mere passive filters, and we are quite

Fig. 10. If the middle disk of Fig. 8 is removed, no photons emerge from the third disk. The *removal* of a disk from the beam *reduces* the amount of light getting through. This demonstrates that the middle disk was not merely a passive filter, but altered the properties of the photons that emerged from the first disk, to make it possible for them to pass through the third disk.

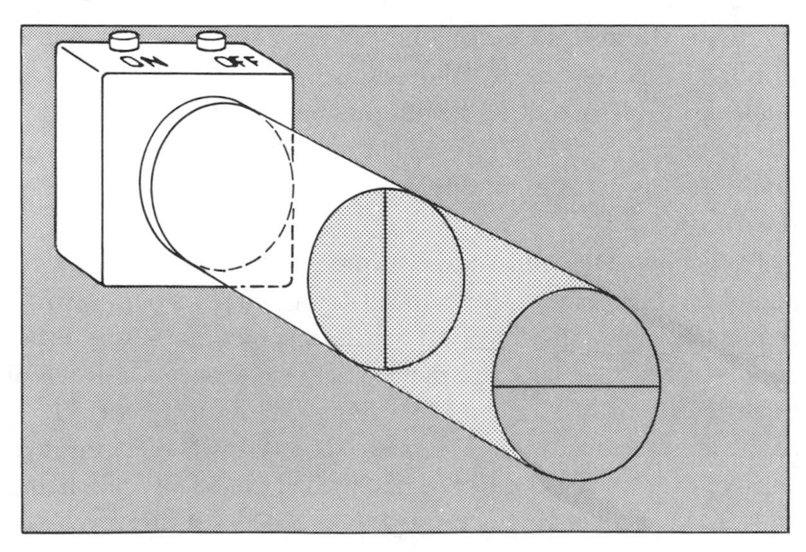

unprepared for the possibility that the middle disk can actively alter the nature of the light that passes through it. This, of course, is precisely why the light can pass through a third disk that would have blocked it had the middle one not been there. Once we recognize this possibility, the behavior is not at all mysterious, though for most people, like a good magic trick it remains very striking to observe, no matter how long or how well you have understood the explanation.

Do photons nevertheless have definite polarizations along all directions?

Even though a disk can alter the polarization of the photons passing through it, each photon might still have a definite polarization along every direction prior to its encounter with a disk and then, after passing through, another definite polarization along every direction. If this were so, the disruption a disk imposed on the photons passing through it would consist of impressing on each a new set of polarizations for all directions.

The question of whether or not a photon has a definite polarization along every direction would now appear to be unanswerable. If we tentatively adopt the view that it does have, then we come up against the fact that the only way to determine the polarization in any given direction is to see whether or not the photon gets through a disk oriented in that direction. But every time we do this the disk can alter the polarization of the photon in all directions but its own. There is therefore no way to test a proposed assignment of polarizations to a photon in more than a single direction and therefore apparently no way to settle the question. Under these circumstances, why worry about it?

This is precisely the point of Pauli's remark that one should not rack one's brains about whether something one cannot know anything about exists all the same. Whether a photon has a polarization along two directions at once is just such a worry. Because determining the polarization along the first direction disrupts the polarization along the second, we can never learn the values of both together. Speculating about whether they nevertheless both exist would appear to be futile.

The Einstein–Podolsky–Rosen experiment

The clever thing about the Einstein–Podolsky–Rosen experiment is that it seems to offer a way around this objection, providing a convincing argument that a photon really does have polarizations in all directions, even if you cannot measure the polarization along one without disrupting it along all the others. To do the Einstein–Podolsky–Rosen experiment with photons, one must produce pairs of photons in a very special way. In the Paris experiments this is done by shining light from a laser on some atoms – calcium atoms (a matter of no consequence for any of the points I wish to make here, mentioned only to emphasize that this is a real experiment done with real materials). The laser knocks some of the atoms into what is called an "excited state," in which condition the atom has more energy than it is necessary for a calcium atom to possess. An atom does not stay for long in an excited state but returns to its state of lowest energy, disposing of the excess energy by emitting photons that carry it off. The laser in these experiments is tuned to knock the atoms into a particular excited state for which two of the subsequently emitted photons reveal some very simple regularities when the behavior of each is tested with polarizing disks. These regularities are straightforwardly predicted by the quantum theory – a prediction that the actual experiment confirms.

Suppose we intercept the path of each photon from such a pair with disks, both oriented in the same direction α (Fig. 11(a)). It is found that invariably either both photons get through their disks, or both are blocked. Whether they pass or not is entirely random – half the time they get through and half the time they don't. But whatever one photon does, the other invariably does the same. This happens whatever the choice of direction α, as long as both disks have the same orientation.

(A technical problem: One knows that a disk has let a photon

Fig. 11. The Einstein–Podolsky–Rosen experiment. (a) Two photons emerge from a calcium atom heading for two disks oriented in the same direction. Invariably either both photons pass through the disks or both photons are blocked. (b) The calcium atom can be placed closer to one disk than the other so that photon #1 gets through its disk before photon #2 has arrived at its disk. Since #1 did get through, so will #2. (c) An extension of the experiment. Photon #1 has passed through a disk oriented along direction α and photon #2 is approaching a disk oriented along β. If photon #2 is really polarized along α (as the passage of photon #1 through its disk and past experience of the experiment strongly suggest) then the probability of #2 passing through its disk is just $\cos^2(\alpha - \beta)$, which is $\frac{1}{4}$ when the angle between α and β is 60°.

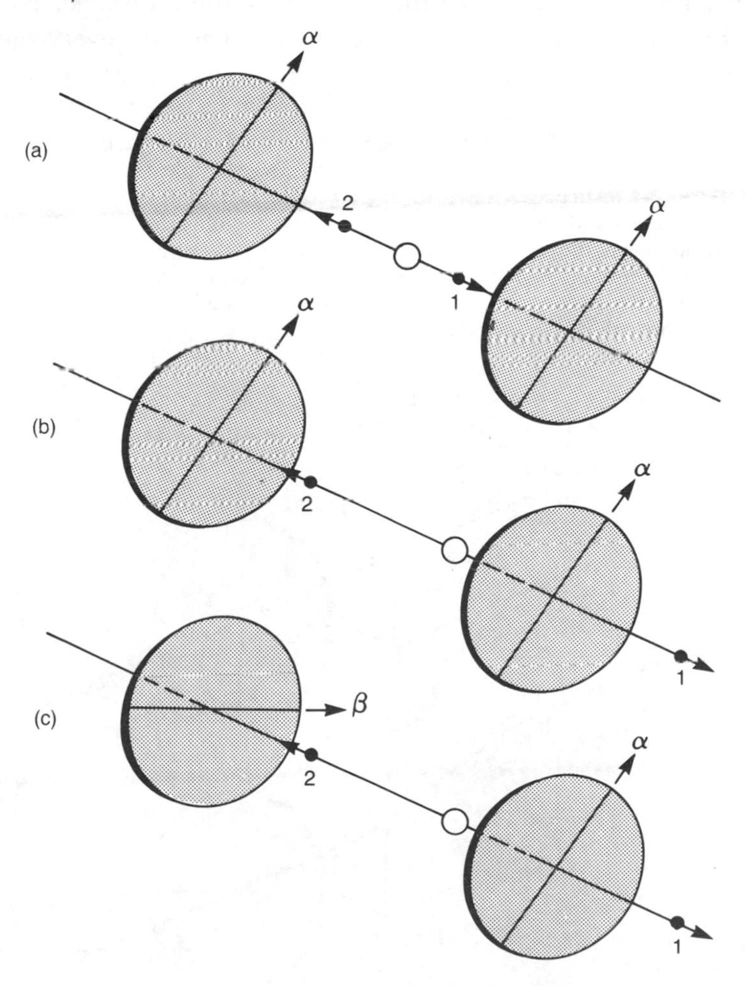

through by detecting the photon on the far side, but how does one know that a disk has blocked a photon without detecting it first on the near side, thereby running the risk of disrupting the photon before it reaches the disk? A technical solution, which is what the actual experiments do: Use an improved version of the disk in which photons of one polarization pass directly through while photons of the perpendicular polarization are no longer blocked by the disk but instead have their trajectories shifted to the side before they pass, as illustrated in Fig. 12. Many naturally occurring crystals can perform this trick. If you want to keep this refinement in mind, read "shifted" for "blocked" everywhere below.)

Fig. 12. A crystal that improves upon the disks. It can be oriented so that (a) light polarized in one direction (by, for example, transmission through a disk) goes straight through it while (b) light polarized in the perpendicular direction suffers a deflection upon going through it. The advantage of the crystal over the disk is that by deflecting photons that a disk would block, the crystal makes it possible to detect photons of either polarization. (With a disk only photons of the polarization that gets through can be detected.)

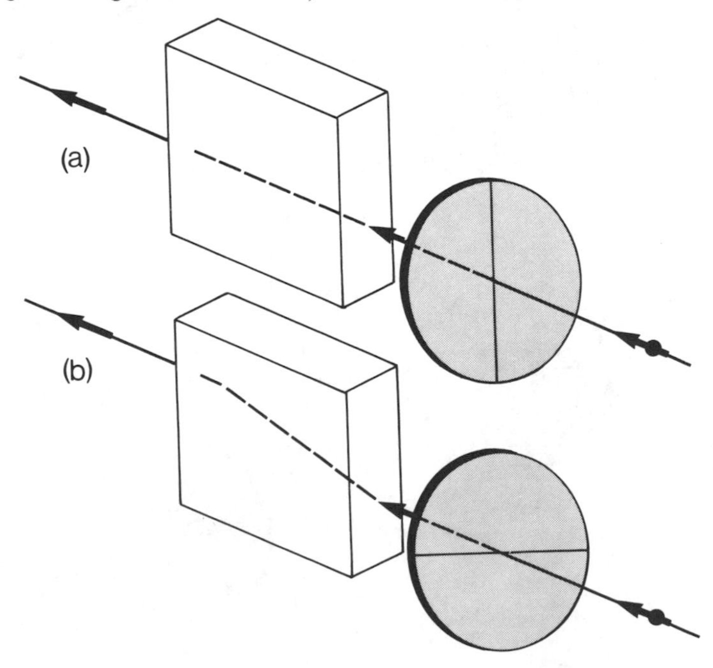

The Einstein–Podolsky–Rosen argument

These facts make perfect sense if both photons have definite polarizations in all directions, and those polarizations are the same for every direction. Indeed, the facts seem to require that both photons have definite polarizations in all directions. The argument goes like this:

Suppose photon #1 gets through its disk, oriented along an arbitrary direction α, before photon #2 reaches its disk, also oriented along α (Fig. 11(b)). From our previous experience with the last thousand (or million, or billion) such photon pairs, we can conclude with confidence that photon #2 is the kind that invariably passes through a disk oriented along α. But to assert that photon #2 is certain to pass through a disk oriented along α is to assert that photon #2 has a definite polarization along α. If, on the contrary, photon #1 fails to get through its disk oriented along α, then we can confidently conclude that photon #2 will also be blocked by its disk oriented along α, which again means that photon #2 has a definite polarization along α.

Please note that I have engaged here in a small amount of sleight of hand. According to our definition, a photon has a definite polarization along a direction α if it has just emerged from a disk oriented along, or perpendicular to, α. One can then confidently predict what such a photon will do at a subsequent disk oriented along α. But it is only on the basis of this secondary property that I concluded that photon #2, which has not emerged from any disk, nevertheless has a definite polarization along α. We shall return to this concern in the next subsection and alleviate it by noting that photon #2 also has all the other secondary properties we demand of a photon with a definite polarization along α.

Our conclusion that photon #2 has a definite polarization along α, even before it encounters the disk along α that tests for that polarization, applies whether or not photon #1 gets through its disk along α. Since photon #1 either gets through or is blocked – there is no third alternative – we conclude that regardless of how photon #1 behaves at its disk, photon #2 has a definite polarization along α. Thus, in the absence of "spooky actions at a distance" – actions that would somehow alter the character of photon #2 as a result of what we do to photon #1 – there is really no reason to subject photon #1 to the disturbing encounter with a disk at all, unless we want to find out the actual value (along α or perpendicular to α) of the definite polarization of #2.

We could, of course, have placed the disks so that photon #2 reached its disk before photon #1. In that case we would have concluded, in exactly the same way, that photon #1 must have had a definite polarization along α right from the start. Thus, both photons must have had definite polarizations along α. Furthermore, since the conclusion that one photon has a definite polarization along the direction α does not require an actual measurement of the polarization of the other along that direction (again, in the absence of spooky connections), and since *not* measuring polarization along a direction α is the same as *not* measuring it along any other direction β, we are led to conclude that both photons must have definite polarizations along every conceivable direction.

A slightly different route to this same conclusion is very reminiscent of the reasoning we applied to the device. Whether we test photon #1 with a disk along directions α, β, γ, ... a subsequent test of photon #2 along the same direction will give the same result: #2 will pass its disk if and only if #1 has passed. But we can wait until the very last moment before deciding how to orient the disk for photon #1. Photon #2 must be prepared to give the same result as photon #1, regardless of which choice we make, and must therefore be carrying instructions that specify what it is to do for all possible directions of its disk. These instructions are simply the definite polarizations #2 must have along the directions α, β, γ, ...

Strengthening the Einstein–Podolsky–Rosen argument

As noted above, we have, strictly speaking, gone too far in reaching conclusions about photons possessing definite polarizations, since our definition of "definite polarization" only asserted that a photon had such a definite polarization along a direction if it had just emerged from a disk oriented along, or perpendicular to, that direction. But the photons we now wish to endow with definite polarizations did not emerge from any such disks. What we do know about them is that they share with photons of a definite polarization a very important property: namely, we can predict with absolute certainty, in advance of their arriving at a disk, whether or not they will pass through the disk (which we do by first determining whether the other photon of the pair passes or does not pass through another disk oriented in the same direction). That, however, is only one of many properties of photons with a definite polarization. In particular, if we confront a photon polarized along a specified direction with a disk oriented along some different direction, there are definite rules about the fraction of times it will get through the second disk. But if photon #2 is really polarized along the given direction these rules should also be obeyed. And they are.

Suppose we intercept photon #1 with a disk oriented along α and intercept photon #2 with a disk oriented along a different direction, β (Fig. 11(c)). Suppose photon #1 reaches its disk first and passes through. We can conclude that photon #2 would also pass through a disk oriented along α, were one placed in its path, but the disk we have actually placed in its path is oriented along β. If photon #2 is really polarized along α, then we know the probability that it will pass through the disk along β. If, to take the example of most importance for understanding the device of Section II, the angle between α and β is 60°, then that probability is 25%. Therefore if the passage of photon #1 through the first disk really establishes that photon #2 is also polarized along α, then photon #2 should get through the second disk just 25% of the time. And it does. Whatever the angle between the two disks, photon #2 always behaves at its disk precisely as if it were a photon polarized in the manner revealed by what photon #1 does at its disk.

The device revealed; the Einstein–Podolsky–Rosen argument demolished

These last facts appear to add powerful additional support to the proposition that a photon possesses a definite polarization along any direction whatever, even though that polarization can be utterly disrupted by an attempt to ascertain its value along even a single direction. But appearances are deceptive. When examined more carefully, the facts we have just described entirely undermine this conclusion.

How the conclusion is refuted was, in fact, explored in great detail in Section II. To see this it is only necessary to note (at last) that the experiment we have built up to is nothing but our

Fig. 13. How the detectors in the device actually work (oversimplified, but basically correct). A photon (black blob) entering the circular entrance horn of a detector encounters a disk. The three settings of the switch on a detector correspond to three different orientations of the disk that are 60° apart. Disks in these three orientations are shown as they appear to the photon in the three insets above the detector. The light flashes green if the photon gets through a disk and red if it is blocked.

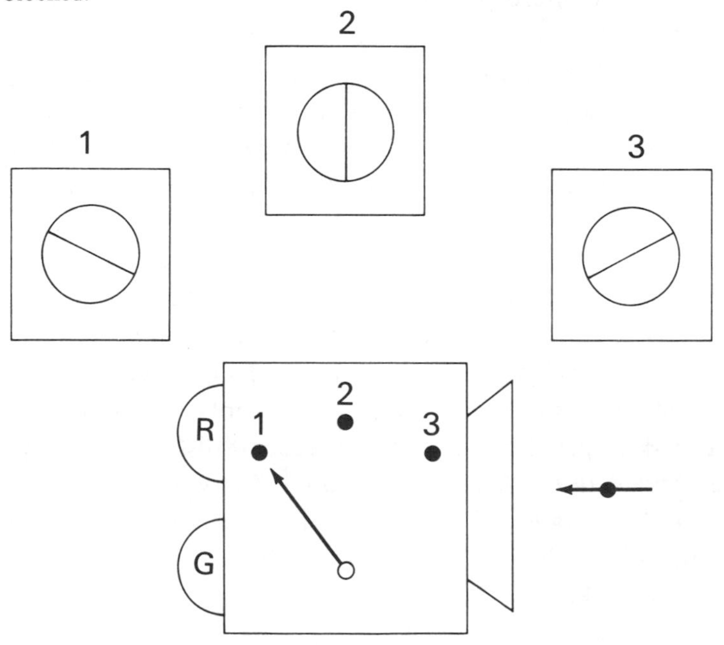

amazing device in a not very obscure disguise. To construct our device out of the Einstein–Podolsky–Rosen experiment, let the three settings of the switches (1, 2, 3) at each detector correspond to three different orientations of the disks, each at a 60° angle to the other two (Fig. 13). Employ the convention used by traffic lights, letting the light at a detector flash green whenever a photon passes through its disk, and red whenever a photon is blocked. The first feature of the data produced by the device is now easy to account for. The lights always flash the same color when the switches have the same setting because the detectors are then measuring the polarizations of the two photons along the same direction, and the photons always reveal the same polarization whatever that direction is.

To see that the second feature must also hold – that the same color flashes as often as different ones when all runs are examined without regard to the switch settings – note that in any run in which the switches have different settings (namely, the six types of runs – 12, 13, 21, 23, 31, and 32) the two disks will be at 60° to one another, the polarizations will be the same only 25% of the time, and therefore the lights will flash the same color only 25% of the time. Consider then what happens over a great many runs in which the switches are set at random. In $\frac{1}{3}$ of those runs the switches end up with the same setting and the lights flash the same color. In the remaining $\frac{2}{3}$ of the runs the switches end up with different settings and the lights flash the same color only 25% of the time. Since 25% of $\frac{2}{3}$ is $\frac{1}{6}$, this gives us an additional $\frac{1}{6}$ of all runs in which the lights flash the same color. The total fraction of all runs in which the lights flash the same color is therefore $\frac{1}{3}+\frac{1}{6}=\frac{1}{2}$: in all runs taken together, the lights flash the same color half the time (and therefore different colors the other half).

So now you know what is in the device. The source C contains a laser and some calcium atoms. When the button on C is pushed, windows on either side of C open up to release an appropriately correlated pair of photons produced by shining the laser light on the calcium atoms. The detectors A and B each measure photon polarization. The settings of the switches determine along which of three possible directions at 60° to one another polarization is measured. For a given switch setting a detector flashes green if the photon is polarized along the corresponding direction, and red if it is polarized perpendicular to that direction.

The instruction sets we seemed forced to infer from the first feature of the data are thus the polarizations of each photon along three directions at 60°, and our subsequent demonstration that instruction sets are inconsistent with the second feature of the data establishes that the photons cannot have polarizations along all three of those directions. Irresistible as it seems to be, the Einstein–Podolsky–Rosen conclusion is wrong.

Whether or not you have followed this discussion of photon polarization in detail, please note that the brief paragraph above explaining how the device works makes no use of any connections between the three parts of the device other than the fact that the lights flash as a result of the passage of photons from the source to the detector. There are no connections between the parts of the device.

IV. Living with the device

Where does this leave us? We have a phenomenon (light always flashing the same color when the switches have the same setting) that cries out for explanation, but the only explanation that seems at all satisfactory is easily shown to be utterly unacceptable. In this concluding section I survey some of the possible responses to this strange state of affairs.

1. Deny that the device is possible

This position is taken by a very tiny minority of physicists, unnoticed by most of their colleagues and viewed by most of those who do notice as something of a lunatic fringe. To hold their position it is necessary either to maintain that the quantum theory, in spite of its spectacular sixty-year record of success, is fundamentally unsound in one of its most elementary applications, or that there are new, hitherto unknown forces in the world that interfere with a pair of separating Einstein–Podolsky–Rosen particles in such a way as to prevent the device from functioning as I described in Section II.

It is also necessary for such people to find flaws in the apparently successful efforts to confirm the quantum theoretic predictions by constructing versions of the device with photons and polarization detectors. Indeed, if you look with near paranoid suspicion at such experiments, loopholes can be discerned. Probably the most conspicuous one is this: detecting the presence of a single photon is not an easy feat to accomplish, and the detectors do not always fire when a photon enters. In fact, more often than not, they fail to fire.

The data collected in the actual experiments come only from those runs in which both detectors actually flash (and flash close enough together in time to establish that both photons were indeed emitted by the same calcium atom). To distinguish those successful runs from those in which a detector fails to flash requires a comparison of what happens at each detector – i.e., a connection, notwithstanding my earlier insistence *ad nauseam* that no such thing exists. This particular connection is placed in the category of "irrelevant connections" by the assumption that those photon pairs, both members of which are detected, are typical of all the other photon pairs. This is highly plausible, because everything we know about how photon detectors actually work indicates that whether a given detector fires or not is an entirely random matter. Nevertheless, if one were stubborn, one could attribute the failure of a detector to fire not to the apparently well-understood problems in building good photon detectors but to certain new, currently poorly understood properties inherent in the photon itself. By pursuing this strategy we can construct a simple "explanation" for the correlations produced by the device.

Suppose particles carry instruction sets like RRG, GRG, etc., but in addition to an "R" or a "G" an instruction set can also carry a third type of entry: "X." When a particle carrying X as the instruction for a given switch setting reaches a detector with that particular setting, the detector does not fire at all, and therefore such a run does not result in any recorded data. In this proposed explanation, the detectors fail to fire not because of their inherent inefficiencies but because the particles themselves can tell the detectors not to fire. They are thus still perfect detectors, but they (and the photons) are conspiring to simulate imperfect ones.

Here is one example[19] of how, given such a peculiar interpretation of the failure of the detectors to fire, one could account for all the data produced by the device. In each run of the experiment let one of the particles carry one of the six instruction sets, RRG, RGR, GRR, GGR, GRG, or RGG (selected at random), and let the other particle carry exactly the same set except that one occurrence of the color that appears twice is replaced by an X. If, for example, one particle carried the instruction set RGR, then the other would carry either XGR or RGX. The two possible positions of the X, as well as which particle carried the instruction with the X, can vary randomly from one run to the next. Under these conditions a third of the runs would have to be discarded, because one or the other of the detectors had its switch set to the position of the X instruction and therefore failed to flash. In the remaining runs the lights would have to flash the same color when the switches had the same settings, because the instruction sets carried by the paired particles agreed in the two positions where neither carried an X. Furthermore, if we examine any of the pairs of instruction sets that can be issued – for example GRR (for the particle heading toward A) and GRX (for the one going to B) – we find that out of the six switch settings for which it produces a pair of flashes (11, 12, 21, 22, 31, 32), half of them give flashes of the same color (11, 22, 32) and half, flashes of different colors (12, 21, 31). Thus the second feature of the data, which demolished the hypothesis of instruction sets, can also be accounted for by using these generalized instruction sets which control when the detectors fail to fire.

So if you are stubborn, there *is* a way to explain the observed data with instruction sets, one that can only be ruled out (assuming our current ideas about photons and their detectors are in fact correct) by designing photon detectors so efficient as to leave no room for such a loophole. The explanation does, however, have a disturbingly conspiratorial quality to it and requires some bizarre coincidences. Einstein once said that God is subtle but not malicious. If the explanation for the behavior of the device lies in the fact that the photons are exploiting our failure to understand detector inefficiencies in just such a way as to mimic the predictions of the quantum theory in those runs that are detected, then if not malicious, God is surely a practical joker, and not a very subtle one. If a valid explanation were nevertheless to be found along these lines, it would, of course, not only entail the collapse of our present understanding of photon detectors but would also bring the entire structure of the quantum theory down in ruins. Few expect this to happen.

2. Embrace action at a distance

If one believes that the device does behave as advertised, and insists on finding a mechanism to explain its behavior, then it seems necessary to invoke some kind of action at a distance. For example, one can assert that when one particle of a pair passes into detector A with its switch set to 1 and the green light flashes, this *converts* the other faraway particle to one that will cause the green light to flash at detector B, if B's switch is set to 1. Details of how the conversion takes place are not supplied by existing physical theory, but that it has to be conversion – a change in the character of the particle – seems unavoidable:

If we can produce a group of particles that invariably result in a detector with its switch set to 1 flashing green (or red) let us say that those particles are "1-colored" green (or red), and similarly for 2- and 3-colors. (If you feel at home with photons and their polarizations, then note that for a photon to be 1-colored green means simply that it is polarized along the direction corresponding to switch setting #1; 1-colored red means polarized in the perpendicular direction.) Suppose we arrange things so the detector A always flashes before B (by bringing A close to the source C and moving B far away), and suppose we set the switch at

A to position 1. Then the group of particles approaching B whose companion particles caused A to flash green will be 1-colored green, because after the arrival of every one of them B will flash green, while the other particles arriving at B (companions of particles that caused A to flash red) will be 1-colored red. Every particle approaching B will thus have a definite 1-color.

Now the particles at B could not have had their definite 1-colors *before* their companion particles entered detector A, because *before* the particles entered detector A it might well have been decided to set A's switch to 2 or 3 instead of 1. "Before a particle entered detector A with its switch set to 1" is indistinguishable from "Before a particle entered detector A with its switch set to 2 [or 3]." If the particles at B had their 1-colors before anything happened at A, then they must also have had 2- and 3-colors. But to possess definite 1-, 2-, and 3-colors is to carry an instruction set. And instruction sets, as we have seen, cannot exist.

Each particle at B can therefore have had its definite 1-color only *after* its partner entered detector A with its switch set to 1. At that point it would no longer be possible for the partner to enter A with its switch set to 2 or 3 instead, and therefore it would not be possible to endow the particle at B with the other two entries of a complete instruction set. The particle at B must thus have acquired its 1-color by virtue of the faraway detection of its companion particle at A.

If this is action at a distance, it is of a very peculiar variety. For one thing, through the magic of relativity, conditions can be arranged under which the result of a given run – say both switches set to 1 and both lights flashing red – will receive different interpretations from different observers. Some will insist that it was the detections at A that 1-colored the particles at B, while others will maintain that it was the detections at B that 1-colored the particles at A. This certainly undermines one's confidence in the explanatory power of the proposed mechanism.

On the other hand, there turns out to be nothing wrong with this divergence of opinion, because the action at a distance is entirely useless. If a series of runs is conducted with the switch at A fixed in one of its three positions, there is no way to infer from the pattern of colors at B whether one is detecting particles with definite 1-, 2-, or 3-colors: the pattern at B is entirely random, whichever way the switch was set at A. The only manifestation of the switch setting at A having any relation to what happens at B emerges when you compare what happens at both detectors. People at B can make such a comparison only by having information about the run at A sent to them by some second more conventional method (telephone, radio, messenger). Under these conditions the only function of the claim that something was transmitted from A to B appears to be the relief it affords us from a certain intellectual pain; the relief is just as great if one asserts that the transmission was from B to A. But the headache comes roaring back if you contemplate both explanations at once.

There are those who conclude that the lesson of the device is that "spooky actions at a distance" are real – that in acquiring a definite 1-, 2-, or 3-color, the particle at B has changed as a result of the detection of thc particle at A. Others insist that nothing at all changes at B as result of the detection at A; all that happens is that we *know* something more about the particle at B. One is then faced with the question of whether the assertion that a particle has a property is a statement about something inherent in the particle, a statement about our knowledge of the particle, or some subtle combination of the two. These are questions better faced by philosophers – so far without notable illumination – than by physicists. Neither position seems satisfactory, but neither position seems to entail any clear-cut consequences on the basis of which it can be ruled out or confirmed.

3. Beware of reasoning from what might have happened but didn't

A quite different response to the puzzlement raised by the device is to challenge much of the apparently innocent line of thought we pursued in arguing for the existence of instruction sets. At the heart of that argument is the consideration of various possible events, only one of which can actually happen. This is, of course, something we do all the time. The future is not known in perfect detail, we have to be prepared for various contingencies, and we are used to examining a range of possible futures to confirm that we are well prepared for whatever might come next. The argument for instruction sets is precisely of this type: there are nine possible futures facing the particles in any run (the nine possible switch settings). In three of them (11, 22, 33) the lights have to flash the same color. To be prepared for all contingencies, the particles must therefore carry complete and identical instruction sets in every run.

If you look at it closely, this kind of argumentation relies on a requirement of consistency between different possible futures, only one of which can actually happen. Consider, for example, a particular run, say no. 3 871 282, in which detector A ended up set to 1, B to 2, A flashed green and B flashed red (12GR). Our analysis requires that in this particular run, both particles carried either the instruction set GRR or both carried GRG. How do we know that the particle detected at A carried an R in the second position, when the switch at A was actually set to 1? The argument is that if, instead, the switch at A had been set to 2, then since both switches would be at 2, the lights would have flashed the same color. Since the particle detected at B was revealed to have an R in its second position, so must the particle detected at A.

Implicit in this is the assumption that if the switch at A in run no. 3 871 282 had been set differently, then nothing would have changed at B: the light there would still have flashed red. This appears to be nothing but the requirement that there should not be spooky actions at a distance: changing the switch setting at A should not change what happens at B. There are, however, two very different interpretations of such a requirement. The first is statistical: if we do an enormous series of runs with the switch at A set to 1 and then a second series with the switch at A set to 2, the data produced at B will have the same character in either case. Nothing in the data at B can enable somebody at B to figure out how the switch at A was set in each of the two series of runs. This is a requirement of physical locality. If it did not hold, one could use the device to signal from A to B. Our device fulfills this locality requirement, since the data produced at B in either of the two such series is a random sequence of Rs and Gs.

In applying this locality requirement to the instruction sets in run no. 3 871 282, however, we give it a very different interpretation. We are now saying that the same color should flash at B, whether run no. 3 871 282 was a 12 run or a 22 run. But at most one of those two possibilities can actually be realized. Does it make any sense to insist that locality requires a consistency between two events, only one of which can actually happen? To be sure, one thinks this way all the time. I know perfectly well that the outcome of tonight's New York Mets baseball game will not depend on whether or not I watch it on television. If I were very circumspect I would have to acknowledge that what this really means is that if my decision to watch a game is entirely uncorrelated with what I know about the game in advance, then the record of the Mets in those games I watch is entirely indistinguishable from their record in those games I don't watch. Given the two records there is no way to tell which is which. But if I were not circumspect, I would also want to say (and deep in my heart truly believe) that what happens in *tonight's* game – not just games in general but the specific particular individual game to be played tonight – cannot depend on whether or not I watch it on television.

The lesson of the device may be that though for most practical (and all non-quantum) applications this is a perfectly harmless and often quite convenient way to think, it is ultimately flawed, and these flaws are dramatically revealed through the operation of the device. My personal hunch is that this comes closest to capturing the correct lesson. To be sure, it is surprising that nature has actually provided us with a *proof* that it is illegitimate to reason in this way from what might have happened but didn't, but once one accepts this prohibition, it is impossible to deduce the existence of instruction sets from the performance of the device. Unfortunately one bothersome question remains:

Why do the lights flash the same color when the switches have the same setting?

4. Stop being bothered

Most physicists, I think it is fair to say, are not bothered. A minority would maintain that this is because the majority simply refuse to think about the problem, but in view of the persistent failure of any new physics to emerge from the puzzle in the half century since Einstein, Podolsky, and Rosen invented it, it is hard to fault their strategy. The mystery of the device is that it presents us with a set of correlations for which there simply is no explanation. The majority would probably deny even this, maintaining that the quantum theory does offer an explanation. That explanation, however, is nothing more than a recipe for how to compute what the correlations are. This computational algorithm is so beautiful and so powerful that it can, in itself, acquire the persuasive character of a complete explanation. I have been told, for example, that the explanation for the correlations is to be found in a conservation law. What this means is that when carrying out the recipe, one must make use of this conservation law; if there were no conservation law, the recipe would not yield the correlations. But if you ask how, in the absence of instruction sets, the conservation law can possibly work, you will be told gently (or roughly, depending on whom you ask) that conservation laws are conservation laws, and it makes no sense to ask how they "work."

This should not necessarily be dismissed as a cowardly refusal to face hard questions. When Newton explained the motion of the planets along elliptical orbits as a consequence of the inverse square law of gravitational attraction by the sun, many of his contemporaries took the view that he had explained nothing, but merely provided a powerful computational algorithm for calculating those orbits. At best, the real question had been shifted from "Why do the planets move along ellipses?" to "Why do the planets behave as if attracted to the sun by an inverse square law of force?" Subsequent generations learned that this was not a fruitful question to ask, and the law of gravity acquired the character of one of the few fundamental irreducible facts on which all the rest of our knowledge is based. Einstein's subsequent revision of Newtonian gravity, some 230 years later, grew out of a very different set of puzzles and would probably, if presented to the skeptics of Newton's time, also have been viewed as more of a calculational recipe than an explanation.

Thus it is important to know what questions to ask. Many of the most profound advances of science have entailed the recognition that certain questions, formerly regarded as fundamental, should not be asked at all. It appears that "Why do the lights flash the same color when the switches have the same setting?" is such a question, and that the real puzzle posed by the device is not how the world can behave in such a manner, but what it is in the way we think about the world that causes us to find such behavior so puzzling. It would be rash, however, to rule out completely the possibility of surprises, and I conclude this essay, as I began it, with a quotation from Feynman, which eloquently expresses the view a prudent person might well adopt:

> We always have had a great deal of difficulty in understanding the world view that quantum mechanics represents. At least I do, because I'm an old enough man that I haven't got to the point that this stuff is obvious to me. Okay, I still get nervous with it.... you know how it always is, every new idea, it takes a generation or two until it becomes obvious that there's no real problem. It has not yet become obvious to me that there's no real problem. I cannot define the real problem, therefore I suspect there's no real problem, but I'm not sure there's no real problem.[20]

Notes and references

1 Richard P. Feynman, *The Character of Physical Law* (Cambridge: MIT Press, 1965), p. 129.
2 Werner Heisenberg, *Physics and Philosophy* (New York: Harper, 1958), p. 42.
3 Christian Møller, unpublished interview in *Sources for History of Quantum Physics*, July 1963.

4 Aage Petersen, *Bulletin of the Atomic Scientists*, September 1963, p. 8.
5 Werner Heisenberg, *Daedelus* **87** (Summer 1958), p. 100.
6 Heisenberg, *Physics and Philosophy*, p. 129.
7 Heisenberg, *Physics and Philosophy*, p. 48.
8 *Discover*, October 1982, p. 69.
9 *Scientific American*, November 1979, p. 158.
10 Karl Przibram, ed., *Letters on Wave Mechanics* (New York: Philosophical Library, 1967), p. 31.
11 Abraham Pais, *Reviews of Modern Physics*, **51**, 907 (1979).
12 Attributed by Max Jammer, *The Philosophy of Quantum Mechanics* (New York: Wiley, 1974), p. 161.
13 Irene Born, trans., *The Born–Einstein Letters* (New York: Walker, 1971), p. 164.
14 *Born–Einstein Letters*, p. 158.
15 *Born–Einstein Letters*, p. 223.
16 *Physical Review* **47**, 777 (1935).
17 *Physics* **1** (1964), p. 195. This journal was in print for only a short time. For a less technical exposition see J. S. Bell, "Bertlemann's socks and the nature of reality," *Speakable and Unspeakable in Quantum Mechanics* (Cambridge: Cambridge University Press, 1987).
18 *Physical Review Letters* **47**, 460 (1981); **49**, 91 (1982); **49**, 1804 (1982).
19 This very nice way of exploiting the X instruction was pointed out to me by Kenneth Brownstein.
20 Richard P. Feynman, "Simulating physics with computers," *International Journal of Theoretical Physics*, **21**, 47 (1982).

13

A bolt from the blue: the Einstein–Podolsky–Rosen paradox

Over fifty years ago Einstein, Podolsky, and Rosen published a striking argument that quantum theory provides only an incomplete description of physical reality – that is, that some elements of physical reality fail to have a counterpart in quantum-theoretical description.[1] What made this a challenge to the quantum-theoretical *Weltanschauung* was the very mild character of the sufficient condition for the reality of a physical quantity on which their argument hinged: "If, without in any way disturbing a system, we can predict with certainty the value of a physical quantity, then there exists an element of physical reality corresponding to this physical quantity." The argument consists in pointing out that it is possible to construct a quantum-mechanical state ψ with the following properties:

1. The state ψ describes two noninteracting systems (I and II). In almost all discussions the two systems are taken to be two particles, and the absence of relevant interaction is built in by taking ψ to assign negligible probability to finding the particles closer together than some macroscopically large distance.

2. By measuring an *observable* A of system I, one can predict with certainty the result of a subsequent measurement of a corresponding observable P of system II.

3. If, instead, one chooses to measure a different observable B of system I, one can predict with certainty the result of a subsequent measurement of a corresponding observable Q of system II.

4. The observables P and Q are represented in quantum theory

by noncommuting operators, which means that they cannot both have definite values. Einstein, Podolsky, and Rosen give an example in which P and Q are the position and momentum of particle II along a given direction, and it is in terms of this example that Bohr's reply is formulated. Subsequently, David Bohm gave a very simple and clearcut example in which P and Q are distinct components of the intrinsic angular momentum of a spinning particle.[2] In recent experiments by Alain Aspect and his colleagues, they are the linear polarization of a photon along distinct nonperpendicular directions.[3]

The route to the EPR conclusion is straightforward. Because the systems do not interact, the measurements on system I can be made "without in any way disturbing" system II. Therefore, the values of P and Q are both elements of reality. But the quantum-mechanical description of reality given by the wave function ψ does not assign definite values to P and Q. Therefore, the wave function does not provide a complete description of reality.

The EPR article appeared in the *Physical Review* on 15 May 1935. According to Léon Rosenfeld,

> this onslaught came down upon as a bolt from the blue. Its effect on Bohr was remarkable ... A new worry could not come at a less propitious time. Yet, as soon as Bohr had heard my report of Einstein's argument, everything else was abandoned ... In great excitement, Bohr immediately started dictating to me the outline of ... a reply. Very soon, however, he became hesitant: "No, this won't do, we must try all over again ... we must make it quite clear ..." So it went on for a while, with growing wonder at the unexpected subtlety of the argument ... The next morning he at once took up the dictation again, and I was struck by a change in the tone of the sentences: there was no trace in them of the previous day's sharp expressions of dissent. As I pointed out to him that he seemed to take a milder view of the case, he smiled: "That's a sign," he said, "that we are beginning to understand the problem."[4]

Bohr's reply appeared in the *Physical Review* five months later, with exactly the same title as that used by Einstein, Podolsky, and

Rosen: "Can quantum-mechanical description of physical reality be considered complete?" He remarks that the EPR argument can

> hardly seem suited to affect the soundness of quantum-mechanical description, which is based on a coherent mathematical formalism covering automatically any procedure of measurement like that indicated. The apparent contradiction in fact discloses only an essential inadequacy of the customary viewpoint of natural philosophy for a rational account of physical phenomena of the type with which we are concerned in quantum mechanics ... A criterion of reality like that proposed by the named authors contains – however cautious its formulation may appear – an essential ambiguity when it is applied to the actual problems with which we are here concerned ...
>
> [This ambiguity] regards the meaning of the expression "without in any way disturbing a system." Of course there is in a case like that just considered no question of a mechanical disturbance of the system under investigation during the last critical stage of the measuring procedure. But even at this stage there is essentially the question of *an influence on the very conditions which define the possible types of predictions regarding the future behavior of the system.* Since these conditions constitute an inherent element of the description of any phenomenon to which the term "physical reality" can be properly attached, we see that the argumentation of the mentioned authors does not justify their conclusion that quantum-mechanical description is essentially incomplete. On the contrary, this description ... may be characterized as a rational utilization of all possibilities of unambiguous interpretation of measurements, compatible with the finite and uncontrollable interaction between the objects and the measuring instruments in the field of quantum theory. In fact, it is only the mutual exclusion of any two experimental procedures, permitting the unambiguous definition of complementary physical quantities, which provides room for new physical laws, the coexistence of which might at first sight appear irreconcilable with the basic principles of science.[5]

At the end of their article, Einstein, Podolsky, and Rosen anticipate some aspects of this reply:

> One could object to this conclusion on the ground that our criterion of reality is not sufficiently restrictive. Indeed, one would not arrive at our conclusion if one insisted that two or more physical quantities can be regarded as simultaneous elements of reality *only when they can be simultaneously measured or predicted.* On this point of view, since either one or the other, but not both simultaneously, of the quantities P and Q can be predicted, they are not simultaneously real. This makes the reality of P and Q depend upon the process of measurement carried out on the first system, which does not disturb the second system in any way. No reasonable definition of reality could be expected to permit this.

One of the central points at issue is what it means to "disturb" something. I learned my quantum metaphysics primarily through the writings of Heisenberg. As I understood it, the unavoidable, uncontrollable disturbances accompanying a measurement were local,"mechanical," and not especially foreign to naive classical intuition (photons bumping into electrons in the course of a position measurement – that sort of thing). When I read the EPR paper (in the late 1950s), it gave me a shock. Bohr's casual extension of Heisenberg's straightforward view of a "disturbance" seemed to me radical and bold. That most physicists were not, apparently shocked at the time – that Bohr was generally and immediately viewed as having once again set things straight – surprised and perplexed me. Until I learned about J. S. Bell's 1964 paper, "On the Einstein–Podolsky–Rosen paradox," I must confess to having been on Einstein's side of the dispute.[6]

Current attitudes toward the EPR argument and Bell's simple but remarkable analysis vary widely. Physicists today are by and large immune to worries about the meaning of quantum mechanics, and quite oblivious, or, if it is brought forcefully to their attention, indifferent to the intense interest the subject still generates among a few. Thus, Abraham Pais, in his biography of Einstein, expresses the prevailing opinion by suggesting that the EPR paper will ultimately be of interest only for the insight it reveals into Einstein's state of mind (through the phrase "no reasonable definition of reality could be expected to permit this").[7] There is a rather different minority view, of which perhaps the most extreme example is Henry Stapp's opinion that Bell's analysis of the EPR paper constitutes "the most profound discovery of science."[8]

Before Bell's paper, the grounds for dismissing the EPR argument were entirely metaphysical. Pauli, for example, tried to explain to Born in 1954 that the only reason for rejecting Einstein's views was that "one should no more rack one's brains about the problem of whether something one cannot know anything about exists all the same than about the ancient question of how many angels are able to sit on the point of a needle. But it seems to me that Einstein's questions are ultimately always of this kind."[9] Bell transformed the issue by showing that the EPR position was inconsistent with the quantitative numerical predictions of the quantum theory and therefore susceptible to direct experimental test.

How can one demonstrate that "something one cannot know anything about" cannot exist? It turns out to be quite simple. Add to the EPR observables *A* and *B* of system I a third observable, *C*, and let the corresponding observables for system II be *P*, *Q*, and *R*. Suppose a measurement of any one of these six observables can have only two outcomes. We can label the outcomes "yes" and "no" and regard the observables as questions. Suppose the pair of observables *A* and *P* invariably yield the same answers, as do the pair *B* and *Q*, and the pair *C* and *R*, so that by measuring the corresponding system I observable "one can predict with certainty the result of a subsequent measurement" of the corresponding system II observable, or vice versa. Einstein, Podolsky, and Rosen would then conclude that the answers to *P*, *Q*, *R*, *A*, *B*, and *C* are all elements of reality, even when it is physically impossible to measure more than a single one of the observables *A*, *B*, and *C*, or more than one of *P*, *Q*, and *R*.

But now suppose that whenever one asked the questions associated with any other pair of observables (that is, the pairs *AQ*, *AR*, *BP*, *BR*, *CQ*, *CP*), the answers invariably differed. The answer to *P* would then always be the same as the answer to *A* but always opposite to the answers to *B* and *C*; thus *B* and *C* would always have the same answer, which would always be opposite to the answer to *A*. On the other hand, since the answer to *Q* would always be opposite to the answers to *A* and *C*, the answers to *A* and *C* would always have to be the same.

Thus, on the one hand *A* and *C* would always have to have different answers, but on the other hand they would always have to have the same answers. The only conclusion, contrary to the EPR conclusion, is that the answers to all six questions cannot all be elements of reality.

Bell's argument differs from this in only one respect. He considers a version (Bohm's) of the EPR experiment in which the state of affairs is exactly as described above except that (in one particularly simple case)[10] the six pairs of questions that fail to always agree do not disagree all the time, but only three times out of four. This is still enough to establish that all the answers cannot exist, though the discussion now acquires a statistical character. The answer to *P* remains always the same as the answer to *A*, but it is now opposite to the answer to *B* and *C* only three-fourths of the time (not necessarily the same three-fourths). Evidently *A* and *C* (or *A* and *B*) can then agree only one-fourth of the time. A little thought reveals that, in addition, *B* and *C* must have the same answers at least half the time. If this reasoning is repeated, starting with a consideration of the answer *Q*, however, the step requiring a little thought now leads to the conclusion that *A* and *C* must have the same answers at least half the time.

We have again arrived at a contradiction: on the one hand *A* and *C* can agree only one-fourth of the time, but on the other hand they must agree at least half the time. The inescapable conclusion is that all six questions cannot have answers.

Thus, if the quantum theory is correct in its quantitative predictions for this type of EPR experiment, then the EPR conclusion is untenable even for people who reject, or fail fully to grasp, the notion of complementarity. Even a neutralist position like Pauli's must be abandoned. That quantum-theoretical predictions are correct in precisely this context has been conclusively demonstrated in the elegant series of experiments by Aspect and his colleagues, mentioned above.

In the Aspect experiments, systems I and II are a pair of photons, emitted by a calcium atom in a radiative cascade after appropriate pumping by lasers. The observables A, B, C, ... (and P, Q, R, ...) are the linear polarization of photon I (and photon II) along various directions – that is, the question, for a given photon and a given direction, is whether the photon is polarized along ("yes") or perpendicular to ("no") that direction. The initial and final atomic states have spherical symmetry, as a consequence of which quantum theory predicts (and experiment confirms) that the photons will behave in the same way if their polarizations are measured along the same directions. But if the two polarizations are measured along directions 120 degrees apart, then quantum theory predicts (and experiment confirms) that the photons will behave in the opposite way three-fourths of the time. (Aspect *et al.* were interested in a somewhat modified version of Bell's argument in which the angles of greatest interest were multiples of 22.5 degrees, but they collected data for many different angles.)

This is a precise realization of the experiment used above to illustrate Bell's analysis, and the fact that the experiment confirmed the quantum-theoretical predictions to within a few percent establishes that the EPR reality criterion is not valid.

There are some remarkable features to these experiments. The two polarization analyzers were placed as far as thirteen meters apart without any noticeable change in the results, thereby effectively eliminating the possibility that the strange quantum correlations might somehow diminish as the distance between I and II grew to macroscopic dimensions. At such a distance it is hard to deny that the polarization measurement of photon I is made "without in any way disturbing" photon II, in the sense of Bohr's "mechanical disturbance of the system under investigation during the last critical stage of the measuring procedure." Indeed, at this large separation, a hypothetical disturbance originating when one photon passed through its analyzer could not reach the other analyzer in time to affect the outcome of the second polarization measurement, even if it traveled at the fastest possible speed (the speed of light).

In the third of the Aspect papers, the feasibility of bizarre conspiracy theories, designed to salvage the EPR reality criterion, is considerably diminished by the use of an ingenious mechanism for the extremely rapid switching of the directions along which the two photon polarizations are measured. The two switching rates are different, uncorrelated, and so high that several changes are made even during the very short time it would take a signal traveling at the speed of light to reach from one analyzer to the other. Thus, if one takes the view that the "reality" of the polarization of photon I along a given direction depends on the choice of directions along which the polarization of photon II is measured, then that "reality" must be transmitted at superluminal speeds. No reasonable definition of reality could be expected to permit this.

Einstein surely would have found these experiments shocking. In 1948 he observed that an entirely appropriate response to the EPR experiment by "those physicists who regard the descriptive methods of quantum mechanics as definitive in principle" would be to

> drop the requirement for the independent existence of physical reality present in different parts of space; they would be justified in pointing out that the quantum theory nowhere makes explicit use of this requirement.
>
> I admit this, but would point out: when I consider the physical phenomena known to me, and especially those which are being so successfully encompassed by quantum mechanics, I still cannot find any fact anywhere which would make it appear likely that[this] requirement will have to be abandoned.[11]

The Aspect experiments provide such facts. They would not have surprised Bohr, but although some physicists today might regard them as no more than an extremely complicated confirmation of Malus's classical law, they surely would have pleased and interested him. Combined with Bell's elementary analysis they provide a simple, direct, and compelling demonstration of complementarity in one of its most dramatic manifestions.

Notes and references

1 Albert Einstein, Boris Podolsky, and Nathan Rosen, "Can quantum-mechanical description of physical reality be considered complete?" *Physical Review* **47** (1935): 777–80.

2 David Bohn, *Quantum Theory* (Englewood Cliffs, N.J.: Prentice-Hall, 1951), pp. 614–19.

3 Alain Aspect, Philippe Grangier, and Gérard Roger, "Experimental tests of realistic local theories via Bell's theorem," *Physical Review Letters* **47** (1981): 460–3; *idem*, "Realization of EPR-Bohm Gedankenexperiment: A new violation of Bell's inequalities," *Physical Review Letters* **49** (1982): 91–4; Alain Aspect, Jean Dalibard, and Gérard Roger, "Experimental test of Bell's inequalities using time-varying analyzers," *Physical Review Letters* **49** (1982): 1804–7.

4 Léon Rosenfeld, in Stefan Rozental, ed., *Niels Bohr: His Life and Work as Seen by His Friends and Colleagues* (Amsterdam: North-Holland, 1967), pp. 114–36.

5 Niels Bohr, "Can quantum-mechanical description of physical reality be considered complete?" *Physical Review* **48** (1935): 696–702.

6 John S. Bell, "On the Einstein–Podolsky–Rosen paradox," *Physics* (New York) **1** (1964): 195–200.

7 Abraham Pais, "*Subtle is the Lord*": *The Science and the Life of Albert Einstein* (New York: Oxford University Press, 1982), p. 455.

8 Henry P. Stapp, "Are superluminal connections necessary?" *Nuovo Cimento* **40B** (1977): 191–204.

9 Albert Einstein, "Physik und Realitat," *Journal of the Franklin Institute* **221** (1936): 313–347; "Physics and reality," ibid., pp. 349–82 (trans. Jean Piccard).

10 See the detailed but quite elementary discussion of this example in N. David Mermin, "Bringing home the atomic world: quantum mysteries for anybody," *American Journal of Physics* **49** (1981): 940–3.

11 Albert Einstein, "Quanten-Mechanik und Wirklichkeit," *Dialectica* 2 (1948): 320–3; reprinted (in English) as "Quantum mechanics and reality," in *The Born–Einstein Letters*, pp. 168–73.

14

The philosophical writings of Niels Bohr

Vol. I:
Atomic theory and the description of nature
Vol. II:
Essays 1933–1957 on atomic physics and human knowledge
Vol. III:
Essays 1958–1962 on atomic physics and human knowledge
Ox Bow Press, Woodbridge Connecticut, 1987

Once I tried to teach some quantum mechanics to a class of law students, philosophers, and art historians. As an advertisement for the course I put together the most sensational quotations I could collect from the most authoritative practitioners of the subject. Heisenberg was a goldmine: "The concept of the objective reality of the elementary particles has thus evaporated..."; "the idea of an objective real world whose smallest parts exist objectively in the same sense as stones or trees exist, independently of whether or not we observe them ... is impossible ...." Feynman did his part too: "I think I can safely say that nobody understands quantum mechanics." But I failed to turn up anything comparable in the writings of Bohr. Others *attributed* spectacular remarks to him, but he seemed to take pains to avoid any hint of the dramatic in his own writings. You don't pack them into your classroom with "The indivisibility of quantum phenomena finds its consequent expression in the circumstance that every definable subdivision would require a change of the experimental arrangement with the appearance of new individual phenomena," or "the wider frame of complementarity directly expresses our position as regards the account of fundamental properties of matter presupposed in classical physical description but outside its scope."

I was therefore on the lookout for nuggets when I sat down to review these three volumes – a reissue of Bohr's collected essays on the revolutionary epistemological character of the quantum theory and on the implications of that revolution for other scientific and non-scientific areas of endeavor (the originals first appeared in 1934, 1958, and 1963.) But the most radical statement I could find in all three books was this: "... physics is to be regarded not so much as the study of something *a priori* given, but rather as the development of methods for ordering and surveying human experience." No nuggets for the nonscientist.

Going to the other extreme, what have these volumes to say to today's professional practitioners of quantum mechanics, who grew up in physics several generations after the revolution, drinking the strange elixir like mother's milk? It really doesn't bother us very much. We know what goes through the two slits: the wave-function goes through, and then subsequently an electron condenses out of its nebulosity onto the screen behind. Nothing to it. Is the wave-function in itself then something real? Of course, because it's sensitive to the presence of both slits. Well then, what about radioactive decay: does every unstable nucleus have a bit of wave-function slowly oozing out of it? No, of course not, we're not supposed to take the wave-function that literally; it just encapsulates what we know about the alpha particle. So then it's what we know about the electron that goes through the two slits? Enough of this idle talk: back to serious things!

These volumes offer little obvious guidance to the modern thinker of such thoughts. Questions like this are not suggested by anything in these pages, whose explicit style and implicit theme is to be extremely circumspect in all utterances, making no commitments and contemplating no puzzles unless they can meaningfully be stated in terms narrow enough to avoid the commitment or squeeze out the puzzlement. Thinking back to a 1937 conversation with Einstein, Bohr remarks "I was strongly reminded of the importance of utmost caution in all questions of terminology and dialectics." Indeed, he even goes on, to warn "against phrases, often found in the physical literature, such as 'disturbing of phenomena by observation' or 'creating physical attributes to atomic objects by measurements'" which are "apt to cause confusion since words like 'phenomena' and 'observations' just as 'attributes' and 'measurements' are used in a way hardly compatible with common language and practical definition."

These are extremely cautious essays. After reading into the third volume one wants to shake the author vigorously and demand that he explain himself further or at least try harder to paraphrase some of his earlier formulations. But they come out virtually the same, decade after decade. There are, to be sure, small changes. The earlier writings (as late as 1927) regularly mention "the feature of irrationality characterizing the quantum postulate"; later on reference to the irrational disappears.

Here is an example of the kind of disappointment I experienced reading these volumes. Typical quantum effects, Bohr notes, resist pictorial representations. Nor should this surprise us, since our ability to construct such representations or explanations in ordinary language was entirely developed in the course of coping with classical phenomena. Bohr repeatedly insists that we must therefore be content with a purely symbolic mathematical algorithm, which connects one classically specified set of conditions to another. This formalism offers no explanation in the customary sense, but by embracing all possible experimental arrangements it demonstrates the logical consistency of the entire scheme, which is all we can demand of it. But I can't help wondering: mathematics was entirely developed in the course of coping with classical phenomena, so is it not astonishing that it should contain precisely the conceptual tools to bridge this unpicturable, unspeakable abyss? Apparently not. In any event the question never arises.

A friend who has thought more about these things than I have remarked to me that if anybody other than Bohr had written essays like these, nobody would take them seriously. Just what is the man getting at? The major recurrent theme is that complementarity is to be found all about us. Position *vs.*

momentum (or the space–time *vs.* the causal description) are joined by justice *vs.* charity, contemplation *vs.* volition, thoughts *vs.* sentiments, and (perhaps) even humor *vs.* seriousness. What is common to all these dichotomies is never made explicit, but I *think* the lesson Bohr wants us to learn is that we should *always* be wary of seeking a unified description of phenomena which are only manifested under mutually exclusive conditions. (Except, perhaps, in the form of a symbolic mathematical algorithm?) The answer to those really hard questions is that they should not be asked – what Einstein called "the Bohr-Heisenberg tranquilizing philosophy."

But against the sometimes maddening frustration brought about by a study of these ponderous essays is the indisputable fact that nobody has succeeded in saying anything manifestly better in the sixty years since Bohr started talking about complementarity. How he could have known that they would fail, right from the start, is yet another puzzle. As a philosopher Bohr was either one of the great visionary figures of all time, or merely the only person courageous enough to confront head on, whether or not successfully, the most imponderable mystery we have yet unearthed.

15

The great quantum muddle

Review of Karl R. Popper: *The Open Universe* and *Quantum Theory and the Schism in Physics*. Rowan and Littlefield, Totowa, New Jersey, 1982.

These are volumes II and III of Sir Karl Popper's hitherto unpublished *Postscript* to the *Logic of Scientific Discovery*. In addition to the *Postscript* itself, which was largely completed twenty years after *L.S.D.* in 1954, about a third of each volume is devoted to more recent writings. Volume II, *The Open Universe*, subtitled "An argument for Indeterminism", has three "addenda": a reprint of Popper's 1973 essay, "Indeterminism is not enough", a revised version of his 1974 "Scientific Reduction and the essential incompleteness of all science", and a brief essay, "Further remarks on reduction", written in 1981. Volume III, *Quantum Theory and the Schism in Physics*, has a lengthy 1982 preface, and contains a revised and expanded version of Popper's 1967 "Quantum mechanics without the observer".

It was not Popper's original intent that the work should appear in separate volumes. Volume II, though primarily about classical physics, bears importantly on his views of the quantum theory, and further discussions of classical concepts (notably entropy and irreversibility) are in Volume III. I shall therefore discuss the two books as a single coherent work. (Volume I, *Realism and the Aim of Science*, was unavailable at the time this review was written.)

Popper argues with wit, ingenuity, and passion, that the prevailing "Copenhagen" interpretation of the quantum theory is a mistake. Heisenberg, he says, "led a generation of physicists to accept the absurd view that one can learn from quantum mechanics that 'objective reality has evaporated'" (III p. 9).

According to Popper, there really are particles out there, not "wavicles", and each particle has both a position and a momentum. The "transition from the possible to the actual as a consequence of our measurements", one of the hallmarks of quantum behavior, is, in his view, one more of those "gratuitous quantum mysteries and horrors" which collectively constitute "the great quantum muddle" (III p. 133).

These volumes put forth a theory, developed since *L.S.D.*, of how physicists got themselves into such a mess. Their first mistake was to swallow classical Laplacean determinism, "a most unconvincing and unattractive view; and ... a doubtful argument" (II p. 124). Popper attacks from many independent directions the position, held by virtually all physicists, that classical physics, in contrast to the quantum theory, is strictly deterministic. There is a neat relativistic formulation: it is impossible to know all the events in the backward light cone of an event in our future; hence we can never know all the relevant initial conditions and prediction is impossible. There is an argument strikingly like those now enjoying a vogue in the new playground for physicists known as "chaos": except in the artificially simple examples that constitute the bulk of most classical mechanics texts, arbitrarily small changes in initial conditions can lead to wildly divergent patterns of behavior so that, again, prediction is impossible. There is a tricky logician's argument: can a machine that predicts the future predict its own predictions?

Misled by their naive determinism, Popper goes on to argue, classical physicists regarded the use of probabilistic reasoning in physical theories as arising only out of their ignorance of the detailed information required for a rigorously deterministic description. Such an association of human ignorance with probabilistic reasoning can lead to carelessly formulated explanations in which, for example, the irreversibility of macroscopic processes is said to find its explanation not in the nature of the processes themselves, but in our ignorance of the detailed conditions under which they take place. On the contrary, Popper engagingly observes, "hot air will continue to escape even if there is nobody in the quad to provide the necessary nescience" (III p. 109).

This confusion of the objective and the subjective was blown to outrageous proportions with the advent of quantum mechanics. For, although it was then generally recognized that the world is not describable by deterministic laws, the subjective view of probability was not abandoned along with the determinism that gave rise to it. But if probability is held to be subjective and the world is inherently probabilistic, then reality and human knowledge become dangerously entangled. The fact that we cannot know both the position and the momentum of a particle (which Popper accepts) gives rise to the assertion that a particle cannot have both a position and a momentum (which he emphatically rejects).

All the muddle and mystery is dispelled once one recognizes the underlying historic error and starts to think correctly about probability. To begin with, there is Popper's "propensity" theory of probability (also developed since *L.S.D.*) that confers on probabilities a degree of objective reality analogous to that conferred on fields of force since the late 19th century – a good position to assume if the aim is to exorcise subjectivism. In addition, he discusses at great length many classical examples – balls rolling down pin boards, random walkers – that demonstrate the difference between absolute and relative or conditional probabilities. In these examples probabilities change discontinuously with a change in our knowledge, even though there is no corresponding discontinuity in the real world. "This", we are told, "is precisely the same thing as the 'reduction of the wave packet'" (III p. 124).

It is here that the cunning and rather appealing line of thought falls apart. To stress an analogy with the behavior of classical conditional probabilities is to fail to come to grips with the most characteristic and peculiar of quantum facts: it is the amplitudes, not the probabilities themselves (which are the squared moduli of the amplitudes), that combine according to the laws of conditional probability. Popper is well aware of the distinction between probabilities and probability amplitudes, as the Editor, W. W. Bartley III, points out in a lengthy footnote (III p. 72–3), but he makes no attempt to counter the objection any physicist would raise at this point, that it renders all of his many examples irrelevant.

All that is strangest about the quantum theory, from two slit diffraction of particles to the Einstein–Podolsky–Rosen paradox, follows from the fact that it is the amplitudes and not the probabilities that superpose. It is precisely because his pinboard analogy breaks down that such abominations to Popper as Bohr's complementarity of the wave and particle descriptions, or Heisenberg's transition from the possible to the actual are forced upon us. To suggest that all that would never have come about if only Bohr and Heisenberg had better understood the nature of conditional probability comes perilously close to missing the entire point.

In view of his insistence that particles really do have both positions and momenta, it is interesting to see how Popper deals with J. S. Bell's 1964 theorem. This shows, in one simple context, that the assertion that incompatible observables all have values prior to the act of measurement has quantitative measurable implications that are demonstrably wrong, if the computational apparatus of quantum mechanics is correct. Popper's original expectation, he tells us, was that in this case the quantitative results of the quantum theoretic computations would be violated by experiment (III p. 25). This comes close to acknowledging that the theory he had provided with realistic objective foundations was, at least in some of its stickier implications, not the quantum theory at all. With the weight of the experimental evidence now supporting quantum mechanics even in the tight spot Bell pushes it into, Popper declares himself ready, if the results hold up, to admit action at a distance rather than abandon classical realism.

It is a pity these results are so recent and therefore dealt with so cursorily, for the kind of action at a distance they induce a tenacious realist to accept is most peculiar. It cannot be exploited to send signals, and seems to serve no purpose at all other than allowing the realist to maintain some sort of realism. Loose talk about action at a distance under such conditions should be subject to the kind of sharp critical scrutiny Popper himself applies so brilliantly to much other loose talk. But these developments have come too late for him to do much more than repeat the conventional responses, though with his own characteristic twist: that here, for the first time, is an experiment distinguishing between Lorentz and Einstein's interpretations of the Lorentz transformation!

Contemporary physicists will find even more startling Popper's repeated insistence that Bohr and Heisenberg originally held that with quantum mechanics "physics has reached the end of the road; that a further breakthrough is no longer possible" (III p. 6). Einstein's attack on the completeness of the quantum theory, we are told, was aimed at such a doctrine, though "few physicists will believe that it was ever held" (III p. 77). According to Popper, the assertion that quantum mechanics was complete originally meant that the quantum theory of electrons and protons provided the final description of matter, a view that was shaken by the discovery of the neutron and positron ("were these not (previously) hidden variables?" III p. 11) and buried for good by the neutrinos and mesons.

The conventional view, of course, is that the dispute was over whether one should seek an explanation for the general framework of the quantum theory in terms of some more detailed underlying theory, somewhat as Langevin's description of Brownian motion as the result of friction and a fluctuating force can be derived from a more detailed molecular description of the individual collisions between molecules of the liquid (described by variables that are "hidden" on the Langevin level) and the Brownian particle. To assert that the discovery of new particles refuted the claim that the quantum mechanical description was complete, is like arguing that Newton did not, after all, explain the basic facts about planetary motion since he failed to anticipate the subsequent discovery of the outer planets.

The fact is that at each new level of discovery since 1935, the broad conceptual apparatus of the quantum theory – the description in terms of probability amplitudes evolving linearly in time – has survived intact, not only inside the nucleus, but even down into the sub-nucleon world of quarks and gluons. There is no evidence that they are not, all of them, subject to the wave–particle duality, capable of possessing a position or a momentum but not both at once. It is remarkable that these newer discoveries, following things down to length scales almost as much smaller than the scale of atoms as atoms are smaller than us, have neither required nor permitted any modifications whatever in the quantum mechanical *Weltanschauung*.

Popper enjoys this process of turning an argument at 180° to the direction we seem ushered along by all the evidence. Has there been "a fully satisfactory reduction of chemistry to quantum mechanics?" he asks (II p. 142). Never mind that the qualitative and quantitative laws of chemical combination are now subject to detailed computation straight from the equations of Schrödinger or Dirac, limited only by the capabilities of human ingenuity and the power of digital computers in the face of intricate problems in applied mathematics. The reduction is incomplete, says Popper, because we do not fully understand the facts of nucleogenesis. But the origins of the atomic nuclei – how they were cooked from hydrogen in the interior of stars – is no more a concern of conventional chemistry than plate techtonics is a branch of astronomy. Did Newton not reduce planetary motion to a problem in mechanics because he failed to account for continental drift?

"I am sure", says Popper, "I shall shock many physicists who, after having reached my fourth or at most my sixth thesis, will stop reading this rubbish" (III p. 46). I am a physicist, and Popper has me well pegged, though I find his view of the quantum theory to be rubbish of a most stimulating kind, all the way to thesis thirteen and beyond. The fact is that although the underlying quantum mechanical view of the world is extraordinarily confusing – Bohr is said to have remarked that if it doesn't make you dizzy then you don't understand it – yet quantum mechanics as a computational tool is entirely straightforward. One can use it all one's life brilliantly and creatively, without ever entertaining a subjectivist thought or denying the reality of anything. The Founding Fathers worried about such things, and I would guess that to many of those contemporary physicists who have been exposed at all to the answers they gave, that whole aspect of the subject has a quaint, somewhat archaic, and distinctly irrelevant character. Popper offers us a vivid, brilliantly offbeat contemporary reminder that the abyss is still very much there, even if *he* is not made dizzy.

Much of his intensity seems fired by what Popper perceives to be the attitude of contemporary physicists.

> The general antirationalist atmosphere which has become a major menace of our time, and which to combat is the duty of every thinker who cares for the traditions of our civilization, has led to a most serious deterioration of the standards of scientific discussion. It is all connected with the difficulties of the theory – or rather, not so much with the difficulties of the theory itself as with the difficulties of the new techniques which threaten to engulf the theory. It started with the brilliant young physicists who gloried in their mastery of their tools and looked down upon us amateurs who had to struggle to understand what they were doing and saying. It became a menace when this attitude hardened into a kind of professional etiquette (III p. 156).

But the problem is not the difficulty of the technical apparatus of the quantum theory, which is mathematically no more formidable than that of celestial mechanics or classical electromagnetism. Nor do physicists worth serious attention try to impress outsiders by substituting "obscurity for profundity" (III p. 157).

The problem is that although the formalism of the quantum theory fits nature like a glove, nobody, not even Bohr or Heisenberg, has ever really understood what it means. The only entirely concise picture the formalism offers of the world prior to an act of measurement is the formalism itself. One can deplore the resulting instrumentalism, positivism, or subjectivism, but such attitudes are not a disease caught from Mach, Laplace, and

Bishop Berkeley. They have been forced upon us by the atomic world which, with what sometimes seems like an almost playful perversity, refuses to conform to any simple pictures we seem capable of inventing, yet faithfully follows some simple mathematical rules.

Physicists who have come to take their subject for granted would do well to suppress their aversion to rubbish and read these volumes, both to refresh their awareness of how bizarre their subject has become, and to test their own grasp of its foundations against Popper's view that what is most marvelously intricate and subtle in the behavior of the atomic world is just a mystery and horror to be dispelled by some clear thinking about probability.

16

What's wrong with this pillow?

Attitudes toward quantum mechanics differ interestingly from one generation of physicists to the next. The first generation are the Founding Fathers, who struggled through the welter of confusing and self contradictory constructions to emerge with the modern theory of the atomic world and supply it with the "Copenhagen interpretation." On the whole they seem to have taken the view that while the theory is extraordinarily strange (Bohr is said to have remarked that if it didn't make you dizzy then you didn't really understand it), the strangeness arises out of some deeply ingrained but invalid modes of thought. Once these are recognized and abandoned the theory makes sense in a perfectly straightforward way. The word "irrational", which appears frequently in Bohr's early writings about the quantum theory, is almost entirely absent from his later essays.

The second generation, those who were students with the Founding Fathers in the early post-revolutionary period, seem firmly – at times even ferociously – committed to the position that there is really nothing peculiar about the quantum world at all. Far from making *bon mots* about dizziness, or the opposite of deep truths being deep truths, they appear to go out of their way to make quantum mechanics sound as boringly ordinary as possible.

The third generation – mine – were born a decade or so after the revolution and learned about the quantum as kids from popular books like George Gamow's. We seem to be much more relaxed about it than the other two. Few of us brood about what it all means, any more than we worry about how really to define mass or time when we use classical mechanics. In contemplative moments some of us think the theory is wonderfully strange, others think it isn't, but we don't hold these views with great passion. Most of us, in fact, feel irritated, bored, or downright uncomfortable when asked to articulate what we *really* think about quantum mechanics.

I'm one of the uncomfortable ones. If I were forced to sum up in one sentence what the Copenhagen interpretation says to me, it would be "Shut up and calculate!" But I won't shut up. I would rather celebrate the strangeness of quantum theory than deny it, because I believe it still has interesting things to teach us about how we think – about how certain powerful but flawed verbal and mental tools we once took for granted, continue to infect our thinking in subtly hidden ways. I don't think anybody, even Bohr, has done an adequate job of extracting these lessons. From this point of view the problem with the second generation's iron-fistedly soothing attitude is that by striving to make quantum mechanics appear so ordinary, so sedately practical, so benignly humdrum, they deprive us of the stimulus for exploring some very intriguing questions about the limitations in how we think and how we are capable of apprehending the world.

I would guess that an unvoiced reason for such efforts to render quantum mechanics uninterestingly bland is the desire to counter the kind of dumb postquantum anti-intellectualism that says that even the physicists now know that everything is uncertain, leading to the disastrous corollary: anything goes. It is indeed important to emphasize to those who would go from quantum mechanics to know-nothingism, that the quantum theory, far from filling us with paralyzing (or liberating) uncertainty, now permits us to make the most accurate quantitative calculations in the history of science. We must certainly speak up against "the general antirationalist atmosphere which has become a major menace of our time, and which to combat is the duty of every thinker who cares for the traditions of our civilization."

On the other hand it's important in combat to shoot at the right target. The above quotation is from Sir Karl Popper and is directed against the writings of Heisenberg and Bohr. Physicists in the second generation certainly have a much better sense of where to direct their fire; but you run the risk in sanitizing the quantum theory to the point that nothing remarkable sticks out above the surface, that if you go inside and look around you won't find anything left to make it hang together any more.

Thus it is a fact about the quantum theory of paramount importance which ought to be emphasized in every popular and semi-popular exposition, that it permits us to calculate measurable quantities with unprecedented precision. But it does not follow from this that statements that the quantum theory is not deterministic but acausal are vast exaggerations – that the theory has little to do with whether or not nature is a game of probability. Yet it has been argued in this context[1] that even radioactive decay – the very paradigm of acausal discontinuous quantum behavior – appears as probabilistic and abrupt only when an inappropriate question is asked: if a particle is in a state of very well defined energy, then it is inappropriate to ask for the exact time of its decay, and the answer is probabilistic only because the question is not appropriate to the experimental situation.

Now to be sure I can, at least in principle, produce at noon a particle that will decay as the clock is striking midnight, provided I make it in some tricky superposition of energy eigenstates for which asking when the particle decays *is* the appropriate question. But that does not mean that the acausality and discontinuity I associate with the beta decay of a free neutron is somehow my fault, stemming from my having asked the wrong question. The argument that the decay is causal and smooth relies on the fact that the quantum state – which incorporates all there is to know about the neutron – changes continuously, without any jumps, and indeed, deterministically, according to Schrödinger's equation. That's fine. Nevertheless, if I put the neutron into a spherical cavity lined with counters, there will be a rather well

[1] Herman Feshbach and Victor F. Weisskopf, *Physics Today*, October, 1988, p. 9.

defined "ping!" at a rather well defined but unpredictable moment. "All there is to know about the neutron" may well be evolving continuously and deterministically, but that little guy in the cavity goes off discontinuously and probabilistically.

Something interestingly puzzling gets lost by insisting that we are confronted with discontinuity and probability only when we ask a foolish question. The puzzle has to do with the nature of the quantum state, and whether it should be viewed as describing the system, or as describing our knowledge of the system, or as some combination of both, or as none of the above because the quantum state is actually nothing more than an ingredient in a mathematical algorithm for computing the results of a well defined experiment. Using the continuous and deterministic evolution of the quantum state to argue against discontinuity and indeterminism in the atomic world makes more or less sense depending on which of these positions you adopt. If indeed it is nothing more than "all there is to know about the system" that changes continuously and deterministically, then this says nothing about whether the world itself can change discontinuously and probabilistically, unless one takes the position that physics is not about the world but only about "all there is to know" about the world, to which I would say "Ping! Thus do I refute you."

If, however, the state describes the system and not just our knowledge of the system, then I somehow have to think of a neutron as continuously and deterministically leaking electron, albeit in a six dimensional configuration space (nine, if you count the antineutrino too). This introduces continuity and determinism. But the "Ping!" is still there, now being induced by the interaction with the surrounding counters. Are the counters asking the wrong question?

Another thing frequently declared by members of the second generation to be resolved by refraining from asking foolish questions is the puzzlement engendered in some by contemplating the Einstein–Podolsky–Rosen experiment.[2] Usually what is

[2] See, for example, Note 1 or Fritz Rohrlich in *Science* **221**, 1251 (1983)

offered in support of this claim is the observation that there is nothing mysterious in the measurement of the spin of one particle being correlated with the probability distribution of the spin of the other, even if the two are far apart, since the two particles originate from a common source. Nobody would quarrel with that, but what many people find mysterious is not the existence of such correlations, but their particular character, which turns out to be utterly inconsistent with some extremely simple and apparently very reasonable ideas about the kinds of correlations it is possible to have between far apart non-interacting systems exclusively as the result of their having once been together in the same place. It may well be that to ask for any explanation of this "unreasonable" character of the correlations is to ask a foolish question. But the question cannot fairly be dismissed as foolish without saying what it is, and making explicit the simple and apparently reasonable ideas that have to be thrown out with it.

My own view on EPR which keeps changing – I offer this month's version – is that barring some unexpected and entirely revolutionary new developments, it is indeed a foolish question to demand an explanation for the correlations beyond that offered by the quantum theory. This explanation states that they are the way they are because that's what the calculation gives. Some explanations may sound more profound than this – saying, for example, that the correlations are a simple consequence of angular momentum conservation – but that is only because they go into a little more detail about what goes into the calculation. There is, however, an interesting non-foolish question: why do many knowledgeable and thoughtful people feel so strongly impelled to ask the foolish one?

My current version of the answer, not very well developed, is that it has something to do with certain deterministic presuppositions that are built into our thought and language at some deep and not very accessible level, that have somehow infected even the way we think about probability distributions. Being of this frame of mind, I am therefore unwilling to be told *both* that the importance of indeterminism in quantum mechanics has been grossly exaggerated *and* that there is nothing peculiar about the EPR correlations. Einstein once wrote to Schrödinger that "the Heisenberg–Bohr tranquilizing philosophy – or religion? – is so delicately contrived that, for the time being, it provides a gentle pillow for the true believer." When I rest my head on a quantum pillow I would like it to be fat and firm; the recently available pillows have been a little too lumpy to soothe me back to sleep.

III.

Relativity

17

Cruel Nature: a relativistic tragicomedy

Cast of characters

A
Friend of A
B
G
Chorus of Relativists

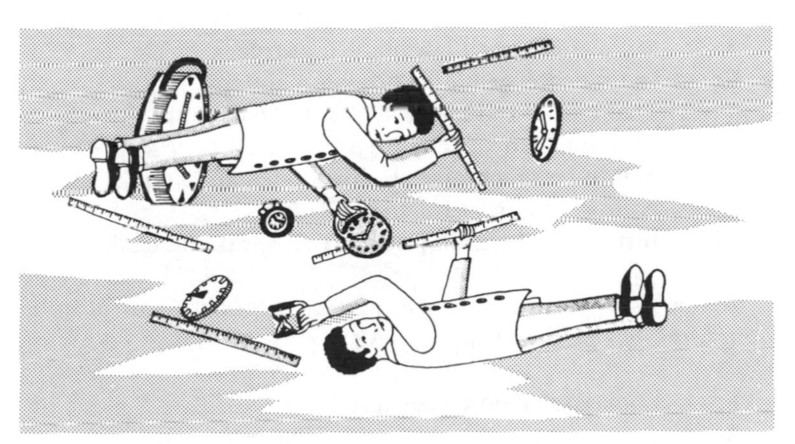

From *Physics Today*, June 1983, p. 9.

A one-act relativistic tragicomedy set in otherwise empty space.

A, surrounded by his clocks and meter sticks, is talking with his friend.

Friend of A: Tell me, good A, is it then truly so
That you are in a state of perfect rest?

A: I am, sir. I move not. My state of rest
Is true and absolute.
F: Is it then so
Your meter sticks do span a meter's length?
A: Not one jot more nor less, sir, I confess,
Provided they maintain their state of rest.
F: How much, sir, in an honest hour's good time
Will these, your clocks, have measured on their dials?
A: Faith, sir, an honest hour! No more, no less,
While they remain with me, at perfect rest.
F: And will each of your clocks, regardless of
The distance 'twixt them, read the same true time
Upon their dials, all in that sweet relation
That does befit fine clocks: Synchronization?
A: This too is so (once more the truth you've guessed!)
Of all my clocks that, with me, are at rest.
F: Your rhymes improve at couplet's grace's expense.
A: Blank verse is not my business. Get thee hence.
F: A thousand pardons, sir! I did but jest
And did not think it would disturb your rest.
A: My rest is perfect, absolute, and true.
F: In that case, gentle A, do you maintain
That clocks and meter sticks that pass you by
With uniform velocity (say v)
Fail to be synchronized, slow down, and shrink
As it is written in Lorentz's Rules?
A: Just so good friend, just so. You speak the truth.

B now floats uniformly into view, seated in the center of an immense network of clocks and meter sticks.

F: Look you! Who comes now?
A: That is Mr. B,
Approaching us with constant speed (say v).
Look how his clocks do fail to synchronize,
Take longer than a second to describe
A second's passage, while his meter sticks
Do shrink in the direction of his motion,
All in accordance with my lovely rules.*
F: Welcome most hearty, B, to A's domain.

* Which in this book are called Lorentz's Rules.

B: Nay, warmest welcome to both you and A
As you progress toward my ancestral home.
F: How fare your many clocks and meter sticks?
B: Now and fore'er, sir, they are just and true.
My clocks are in harmonious accord
And in a second's time do indicate
The passage of a perfect passing second.
My meter sticks extend one meter's length
From end to end.
F: Hear you that, A?
A: I do.
The man has lost his wits. He does not know
That he it is who moves, while I stand still.
Ergo the knave is fully unaware
That all his clocks and meter sticks behave
As it is written in Lorentz's Rules,
Failing to keep true time and span true length
To that extent precise and mathematick
As do my rules require for one who moves
Past me with his velocity.
F: Poor fool!
But now he passes by you and will see
By swift comparison, experimental,
Of his askew equipment with yours true,
That his is deep in error.
A: No, alack!
You overestimate the wisdom of
The man. So deep has he enmeshed himself
In folly, so fully does he deem himself
At rest, that he believes that my Loren-
Tz's Rules describe the sticks and clocks at rest
With me!
F: A double folly's double woe!
But yet methinks there consolation be
In doubleness. The saving point is this:
If to his false-deemed state of rest erroneous
He adds a further concept incorrect
And vile, by his most wrongful application
Of your Lorentz's Rules, which we both know
Describe the strange distortions of things moving
Past him who is at rest (and such are you),
If, as I say (for I have lost the thread

Of my intent) he wrongfully applies
Your special rules, assuming they are his,
Then marry, by this double error gross
(Wrongly to deem himself at rest, and worse,
Wrongly to think that he can use your Rules)
Does he not double chance of contradiction
Which will his fault correct, his mind inform,
When he observes your instruments of measure
So just and true (due to their state of rest)?

A: His second folly does abet his first
And by compounding, save it. Had he thought
Himself at rest and not as well believed
My own Lorentz's Rules, his too to use,
His error, by th' impending confrontation
Of swift advancing B and my true tools
Of space and time, would manifest become
To B himself, forced to this recognition
By contradiction stark and merciless.
Howe'er because he uses my own Rules
As if 'twere he at rest, and I who moved
Along with my true clocks and meter sticks,
The inconsistencies that should inform
His intellect of its sad misconception
And jar it like a ringing clarion call
To certain knowledge of those clear distortions
His many clocks and meter sticks are heir to
By virtue of their motion, he poor fool
Is able to account for in a way
That masks the inconsistencies and bars
Sweet ministering contradiction from
The portals of his mind. He simply blames
The facts that should destroy his sleep dogmatic
On the fictitious shrinkage, slowing down,
And lack of that sweet quality we deem
Most excellent in clocks: Synchronization,
That he in his most vain, deluding use
Of my Lorentz's Rules assigns to my
Most wrongfully malignéd instruments.
To his misfortune, Nature, arch deceiver
So made the world that his delusions, two,
Will learn from this encounter nothing new.
Each doth confirm the other's false surmise.

So was it e'en with C and X and Y.
So shall it be when G and H come by.
I rage against such cruel deceit in vain;
Harsh Nature has decreed it.

F: (*to B, now very close*) Look you, sir:
The clocks and meter sticks of outraged A!
Perceive they argue not the same as yours!

B: Of course they don't: His meter sticks do shrink,
His clocks are slow, nor are they synchronized;
While my sticks measure distance absolute,
My clocks record Time's true and even tread,
Each, though apart, my other clocks do prove.
This is because, quite simply, I don't move.

A: Alas, poor B! Nature conspires against him.

B: Alas, poor A! He thinks that I be mad,
When all too well I know the madness lies
In him. So has it been with Y and C
And X; with G and H, so shall it be.

B passes by A and recedes into the distance.

F: O wicked Nature, to conspire 'gainst B
That all his gross and lamentable follies
Most undetectable thy tricks have rendered.

A: Sadder still, but for delusions twain
He hath a most incisive, cogent brain.
Alas, poor B! And C! And X! and G!

B: (*from afar*) Alas, poor A! And X! And G! And C!

G: (*coming into sight*) Alas, poor A and B! And X! And C!

Chorus of relativists:
Such sorry discord need not be
If Absolutists had more sense:
So right in all their measurements,
So mad in their philosophy.

Dirge: Such Sorry

Lento ma non troppo

18

The amazing many colored Relativity Engine

I. An engine for teaching relativity

I describe below a simple method for teaching the elementary (special) relativistic properties of space and time measurements. This approach was developed and refined during several teachings of a "general education" course on special relativity for nonscientists. Such students are not at home with elementary algebraic manipulations and a direct derivation of the Lorentz transformation along conventional lines is beyond their conceptual powers (nor would they be able to use the equations if they had them). Even a step by step derivation of length contraction, time dilation, and the relativity of simultaneity requires a sophistication with algebraic formalism beyond the attainments of most of these students, and they are entirely distracted from genuine physical and philosophical subtleties by their marginally successful efforts to cope with the mathematics.

One can remove much of this irrelevant but overpowering obscurity by replacing algebra with arithmetic. To make many of the conceptual points it is not necessary to derive everything for general values of v/c. Almost everything is illustrated just as well by special cases, for example $v/c=\frac{3}{5}$. Many students who strain mightily over $(1-v/c)(1+v/c)=1-v^2/c^2$ will swallow $\frac{2}{5}\times\frac{8}{5}=\frac{16}{25}$ with relative ease.

But the arithmetical approach cannot avoid all algebra, and for some students even arithmetic distracts from the real issues. Stimulated by this chronic source of frustration, I was led to design what might be characterized as a hands-on *gedanken* experiment. In this *gedanken* laboratory length contraction, time dilation, relativity of simultaneity, velocity addition, and the like are extracted directly from measurements made with one simulated set of instruments to determine the properties of another such set. No calculations are required beyond the arithmetical manipulation of data. The mutual consistency of all measurements is self-evident. And all the major relativistic effects are directly revealed.

My original idea, in the late 1960s, was to build this pedagogical device into something like a relativistic slide rule – an analog computer, with various sets of gear-linked counter-rotating concentric cylinders representing moving meter sticks and the readings on attached clocks. I imagined the whole thing in brass on a mahogany base, driven by a large ebony-handled crank: a Relativity Engine.

They laughed when I brought them drawings. Nobody would make it for me. So I had to make do with sheets of paper schematically depicting various slices of this gorgeous space–time Relativity Engine.

I got by with pieces of paper throughout the next decade and a half until I noticed that slide rules had vanished, and the campus was suddenly teeming with little computers. The time had come to simulate my Engine on the display screen; students would be able to turn the crank at any of hundreds of conveniently located keyboards. Mahogany and brass were out, but color was in.

This article describes my digital Relativity Engine.

Section II gives a general view of the Engine. It is written with teachers of physics in mind, but it should be intelligible to students with some prior acquaintance with special relativity who might also enjoy playing with the Engine.

Section III is an introduction to the Engine for students who have no acquaintance whatever with special relativity. It is in effect a user's manual. It will also provide teachers with a more detailed picture of how the Engine operates and the pedagogical possibilities it gives rise to. Nowhere in Sec. III do I attempt to put the Engine into the context of an introductory course in special relativity: there are too many different ways one might try, and it seemed to me this was really a matter for the judgment of the individual teacher.

Some suggestions along these lines are given in Sec. IV A, which is again addressed to teachers. Section IV B gives a brief formal analysis of the Engine, which might help those wishing to design Engines of their own with parameters different from mine. Section IV C discusses a few technical features of my computer program and tells interested readers how to get a copy for themselves.

II. Introduction for sophisticates

Place yourself on a space platform somewhere out in the void and imagine two long straight trains (one red, one blue) of identical rockets moving past you and past one another in opposite directions at the same speed. Occupants of the red train are going to measure various properties of the blue one, and vice versa.

The first thing to notice about the Relativity Engine is that it presents both trains from the point of view of a *third* observer (on the space platform), symmetrically situated between the two.[1] Since, except for their directions of motion, the blue and red rocket trains appear entirely equivalent from the third (platform) point of view, it is evident and unavoidable that whatever conclusions the blue observers reach about the red rockets, the red obervers will conclude the same about the blue. Reciprocity is built in.

Each rocket in each train has one porthole window halfway along its length. Alongside the window is the number of the rocket. (The rocket at the front of each train is number 1, the second rocket, number 2, etc.)

Each rocket carries a clock, which can be read by somebody inside the rocket, or just as easily, through the window, by somebody outside. All the clocks on either set of rockets are identical and, in particular, they all run at the same rate. Indeed, as observed from the space platform, there is only one thing at all peculiar about either the rocket trains or the clocks: on neither train are the clocks synchronized. The clock in any rocket (of either train) is ahead of the clock in the rocket just in front of it (and behind the clock in the rocket just behind it). The amount of this disparity is the same from one rocket to the next, and is the same for both rocket trains.

This asynchronization can be viewed in two ways. One can regard it as the ordinary relativistic asynchronization (of order v/c) to be expected in the platform frame, if the clocks are properly synchronized in the frames of the trains on which they reside. Alternatively, however, one can regard the speed of the trains as being so small compared with the speed of light that all relativistic effects are utterly negligible. One can nevertheless imagine the clocks to have been deliberately asynchronized by somebody, either a practical joker, or a professor of physics, out to make a pedagogical point.

The *second* point of view is greatly to be preferred, at least until the students have acquired good *gedanken* experimental technique. One can, at this stage, guiltlessly retain one's Newtonian sense of absolute time, taking the position that the clocks on each train really are asynchronized. *But* – and here is where the fun begins – the people in the rockets do not know that their clocks disagree. On the contrary, they have been lied to, and told that the clocks in each rocket are in perfect agreement with each other. People in different rockets of the same train have no way of communicating with each other to check this – the rockets have no radios, or provisions for space walking from one to the next. In any event, the people on each train have no reason to be suspicious. Having been assured that their clocks are properly synchronized, they will act on this assumption in interpreting such information as they are able to gather about the other train.

Each train has just one way to gather information about the other. When any two rockets are directly opposite each other, then (and only then) people in the window of one rocket can see the serial number and clock of the opposite rocket. Each rocket is equipped with a camera that can record the moment when its window faces the window of a rocket on the opposite train. The resulting photographs (which, of course, contain the same information regardless of which train they are taken from) record the numbers of the two adjacent rockets and the readings of the two clocks they are carrying.

As the two trains pass each other, people in each rocket are free to take photographs whenever the window of a rocket of the other train is outside their own window. When the two trains have completely passed each other the game ends. The rockets land at Edwards Air Force base, and everybody goes off with their photographs. The people from the blue rockets take their pictures to a conference center in Houston, to discuss what they can conclude about the red train. The people from the red rockets retire to a motel near Cape Canaveral to study the blue train from their own photographs.

Since whenever people on both of two adjacent rockets take photographs of each other's clocks and rockets, the two photographs agree, the data available to one group contains no photographs that contradict the other group's data. Indeed, had each group (somewhat wastefully) chosen to take photographs at all possible opportunities, both groups would have identical data to analyze. This is evident from the operation of the Engine, since the way in which students use it to take photographs is quite symmetric between the two trains.

As it turns out, appropriately chosen *pairs* of photographs provide all of the interesting information. The Relativity Engine permits students to take and study such pairs of photographs.

The trains appear on the upper half of the display screen. Each rocket has its serial number stamped on it. The reading of its clock is displayed next to the rocket. Touching the up-arrow key lets time advance by a small step: Each train moves forward a bit, and the reading of each clock advances. One can undo the passage of time by touching the down-arrow key, which restores the previous configuration. Thus by holding down first the up- and then the down-arrow key one can search forward and backward in time for suitably informative moments in the history of the two rocket trains.

When one finds a moment of particular interest one can make a photograph. This can only be done when the rockets are directly opposite one another (since the information can only be collected when the windows face each other). By touching either the right- or left-arrow key one calls up a pointer between the two trains. Further pressing of these keys moves the pointer to the right or left along the line of rockets until it reaches the pair of rockets whose photograph one wishes to record. Pressing a designated shutter key then records the photograph taken in (either of) those rockets at the lower left on the screen (safely out of the way of the two trains). One can then search further through time (with the up- and down-arrows) and space (moving the pointer with the right- and left-arrows) until one finds another suitable pair of juxtaposed rockets. Pressing the shutter results in a second photograph appearing at the lower right.

The student is invited to use the Engine to take suitable pairs of photographs from which, for example, the people on the red train, assuming their clocks are correctly synchronized, can conclude (a) how fast the blue train is going, (b) the extent of the asynchronization of the blue clocks, (c) how fast the blue clocks are running compared with their own red clocks, or (d) how long the rockets of the blue train are compared with their own red rockets. It is evident from the symmetry of the situation (but students should be invited to verify it directly anyway) that people from the blue train, assuming that *theirs* are the correctly synchronized clocks, will reach the same conclusions about the red train.

One can also call up a meteor that moves parallel to the two trains in the space between them at a third speed, specified in the symmetric platform frame by the user. Should two rockets be positioned for a photograph while the meteor happens to be directly outside of their windows, then the meteor appears in the photograph. One can use pairs of photographs in both of which the meteor appears, to measure its speed from either train. By introducing meteors of appropriate speeds one discovers some interesting things. There is, for example, a special speed with the peculiar property that if a meteor is given that particular speed, then the speeds observers on either train find for the meteor are the same (even though the trains are moving in opposite directions). If one puts down a meteor moving sufficiently faster than the special speed one finds that observers on the two trains can disagree on the order in which two photographs of such a meteor were taken, with interesting consequences if the meteor is behaving in an irreversible manner (for example, burning up). By putting down meteors with a range of other speeds one can try to discover (or verify) the relativistic velocity addition law.

In this way, after several hours of playing with the Engine, students will learn for themselves that under the assumption that their own clocks are synchronized, observers on either train will conclude that clocks on the other run slower than theirs and that rockets on the other are shorter than theirs, with specific numerical measures of the extent of the slowing down and shrinkage. They will produce numerical measures of the speed of one train, according to the other, and measurements of the extent to which the clocks of one are out of synchronization according to the other. They will discover an invariant velocity, and learn that objects moving faster than the invariant velocity that behave in a visibly irreversible manner can provide irrefutable photographic evidence that the clocks on one of the trains cannot be correctly synchronized.

The reason one and the same set of data can lead the passengers on either of the two trains to such apparently contradictory conclusions is entirely evident: the conclusions one chooses to draw from the data depend critically on which set of clocks (blue or red) one assumes to be correctly synchronized. Failure fully to grasp this point is the single most stubborn obstacle to a clear understanding of special relativity. Using the relativity Engine, the point is impossible to miss, since it is evident to the student (who sees things from the vantage point of the space platform) that *neither* set of clocks is correctly synchronized.[2] The red and blue conclusions obviously depend on assuming that either the red or the blue clocks are synchronized, in spite of the evidence that neither set is synchronized (which is, however, available only to those who can see the entire screen and not to the people in Houston or Cape Canaveral, who only have isolated photographs of small parts).

III. Introduction for innocents: a user's manual for the relativity engine

A. Rockets and clocks at various times

Pictured on the screen are two trains of identical rockets. (See Fig. 1 (a).) We will call the upper train (that points to the left) the red train and the lower train (that points to the right) the blue train. (If

Fig. 1. (a) The Relativity Engine in its initial configuration. (On the screen the rockets of each train are solid rocket-shaped blocks, with no internal features other than their serial numbers, and each rocket has a flashing exhaust plume behind it.) (b) The Relativity Engine one time step after the configuration shown in (a). Note that the upper rocket has moved a bit to the left, the lower, to the right, and that each clock on each rocket has advanced by one tick. To get from the picture in (a) to the picture in (b) one touches the up-arrow key; to return from (b) to (a) one touches the down-arrow key.

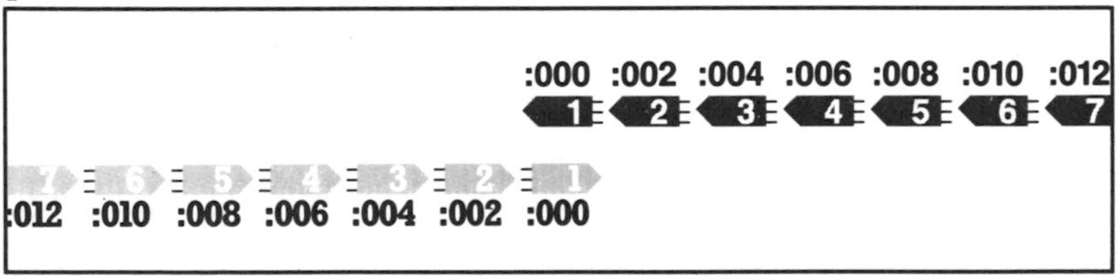

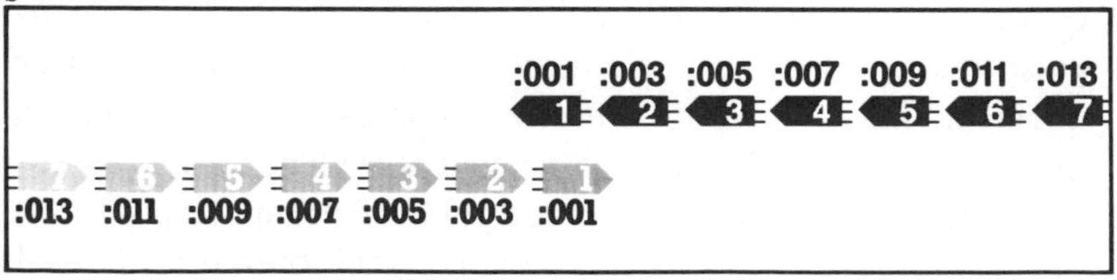

you are using the color version of the Engine, then the red train *is* red and the blue train, blue. If you are using the version that runs on a monochromatic display think politically: The red train goes to the left.)

There are two numbers associated with each rocket of each train:

(1) At the center of every rocket is its identification number. The rocket at the front of each train is numbered 1, and subsequent rockets in each train are numbered 2, 3, 4, . . . , depending on how far down the train they are.

(2) Alongside each rocket (a little above the red rockets and a little below the blue ones) is a three digit number preceded by a colon. The front rockets both show :000, the second rockets in each train show :002, the third, :004, etc. The number alongside a rocket is the reading (in a unit of time we shall call a "tick") of a timer (or a stopwatch – we shall call it simply a clock) that is inside the rocket. If the screen were big enough to show more detailed pictures of each rocket, then this could have been made more vividly evident by showing a window in the middle of each rocket, under which was the rocket's identification number and behind which (i.e., inside the rocket) was a clock showing the number with the colon. Figure 2 shows the big picture that a little picture represents.

Notice that none of the clocks on the red train (we will call them the "red clocks") agree with each other: They all have different readings. These different readings are, however, simply related: as you move toward the rear of the train, the clocks get ahead by *two ticks per rocket.* Thus the clock in rocket 1 reads :000; the clock in rocket 2 reads :002; the clock in rocket 3 reads :004; etc.

The clocks on the blue train (the "blue clocks") are set in exactly the same way: The clock in rocket 1 reads :000, and the readings of clocks in subsequent cars increase by two ticks from each rocket to the one behind it.

Although they are not synchronized with one another, all the clocks on both trains are good clocks of identical construction. In particular, they all run at the same rate. To check this, push the up-arrow key, which produces a new picture, showing what things look like a little later. Notice that two things have changed as a result of pushing the up-arrow (Fig. 1(b)): (a) each train has taken a small step forward (one-sixth of a rocket length, to be precise – a distance we shall define to be one "step") and (b) the reading on every clock has advanced by exactly one tick.

By repeatedly pressing the up-arrow key (or just holding it down if you are in a hurry), you can survey the trains at various times; with each press of the key every clock advances by one tick, and both trains move forward another step. To recapture what things looked like at an earlier time, press the down-arrow, which results in each train moving backward a step and each clock losing a tick.

Thus the up-arrow takes things forward in time, and the down-arrow, backward. It is clear from the sequence of views you get by pressing the arrows that all the clocks on both trains are indeed running at the same rate, and that the trains are simply moving steadily along at the same speed in opposite directions. There is

Fig. 2. A somewhat less schematic picture of what each rocket in a train represents: (a) shows rocket #17 as it appears on the screen, when the clock it carries reads :089 ticks; (b) gives a somewhat more realistic drawing, to emphasize that the clock itself is inside the rocket, but visible through the window just above the serial number.

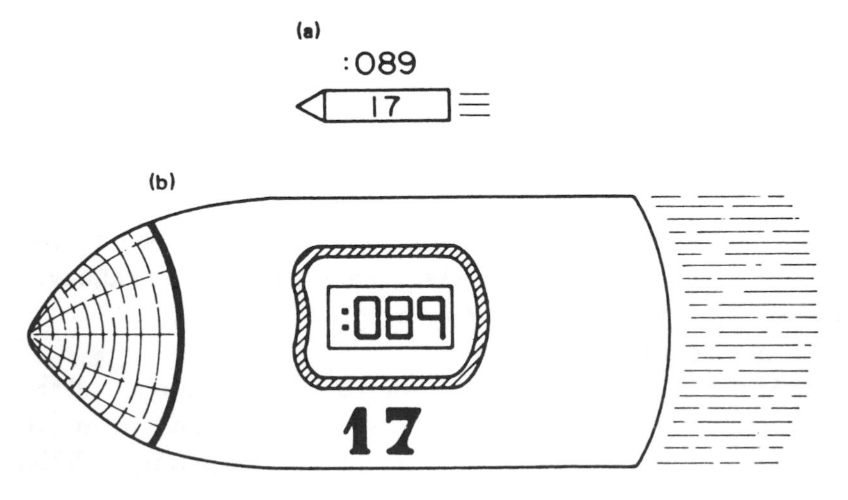

nothing at all peculiar about any of this except for the fact that the clocks are not synchronized.

B. The clock conspiracy

A group of scientists (not on either train) are interested in investigating the kinds of conclusions reached by people using such sets of unsynchronized clocks *if those people are under the false impression that their various clocks actually do agree with each other.* It is these scientists who, unknown to the occupants of either train, have carefully arranged for the clocks in the different rockets of each train to be out of agreement with each other.

The scientists have lied to the occupants of each train about their clocks, telling them that the clocks in the different rockets of their train are in perfect agreement with each other. They have also arranged things to ensure that the occupants of each train cannot learn that they have been lied to. First of all, nobody can move down the train from one rocket to the next, making a direct comparison of the clocks. They are all locked into their own rockets. Furthermore, no mechanism is provided for communicating between different rockets in the train. No radios, no telephones, no leaning out the window and taking a look. The occupants of each rocket are completely isolated from the people in any of the other rockets of the train. The flaming exhaust[3] prevents them from crawling into the nose cone of their own rocket and tapping out a message on the rear of the one in front of them, or vice versa. Any other means of communicating between rockets you might try to invent has been anticipated and made impossible by these clever, determined, and unscrupulous scientists.

The occupants of each train, having no reason to expect deception, believe that the clocks in the various rockets making up their own train are indeed synchronized.

C. Watching the other train

It is a boring journey. There is nothing to do but look out the window, and there is very little to see. With one exception (noted below in Sec. III F), the only time people in either train have anything at all to look at is when the other train passes by. Even this is not very exciting for the occupants of any one rocket, except when a rocket on the other train is directly opposite so that the windows of the two rockets are perfectly lined up. Then, and only then, it is possible for the occupants of each rocket to see the identification number of the other rocket, and note, through the window of the other rocket, what its clock is reading.

At the moment depicted in Fig. 3, for example, the occupants of rocket #5 on the red train are able to look at the clock in rocket #2 of the blue train and verify that at time :023 (according to their own clock) the blue clock in rocket #2 reads :017. In exactly the same way, of course, the occupants of the blue rocket #6 will be noting that at time :025 (according to their own clock) the clock in red rocket #1 reads :015. Shifting our attention up and down the trains, we see that at this same moment other people in other rockets will be reaching other conclusions of the same general character. Note that the occupants of red rocket #8 have nothing to report at this moment, since no blue rocket is opposite them.

There are also moments when nobody on either train has anything of interest to report. In Fig. 4, for example, the rockets are not lined up with each other, and the occupants of adjacent rockets cannot look through each other's windows and determine the reading of the clocks in the other rocket. All the information

Fig. 3. The two trains, shown at a moment when the rockets are lined up directly opposite each other, so that the occupants of any rocket can look through their window and see the serial number and clock reading of the rocket opposite them. Only at such moments of perfect alignment can useful information be collected or can photographs be taken.

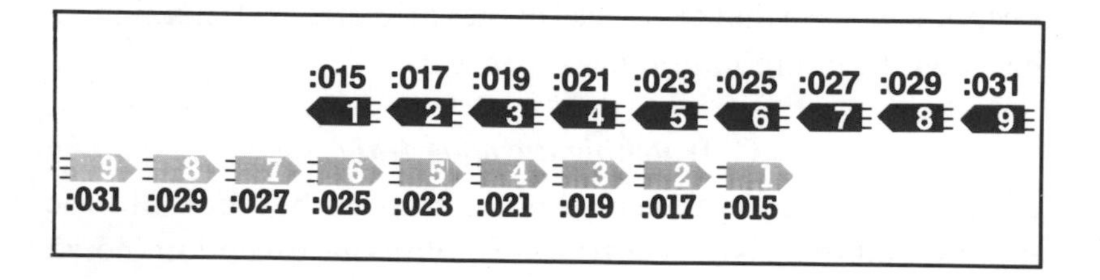

we shall collect and make use of comes from moments of perfect alignment, when each rocket can reveal the reading of its clock to the rocket on the other train that is directly outside its window.

D. Taking pictures of yourself and others

Occupants of the rockets can, whenever they wish, record the information they collect when the window of another rocket appears outside theirs. Thus (Fig. 5) the occupants of blue rocket #8 can note that at "blue time" :041 red rocket #3 is directly opposite and its clock reads :031. This information can be recorded in a sentence, as we have just done, or in some kind of table of data, or, most vividly, in a little picture of the juxtaposed rockets and clocks (Fig. 6) that shows the two rocket numbers and the two clock readings.

Note that such a picture can be used by occupants of either rocket. Thus occupants of the red rocket #3 would conclude

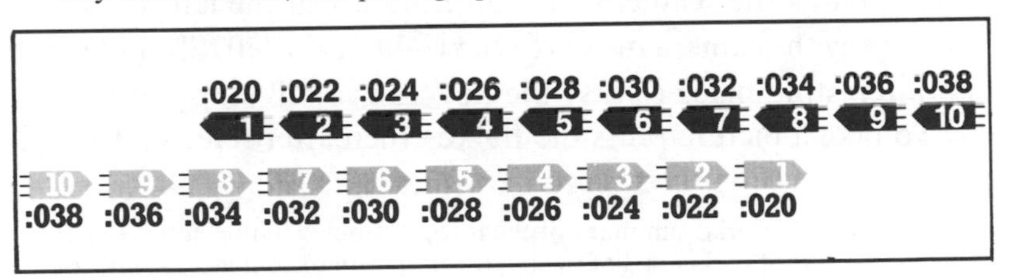

Fig. 4. The two trains, shown at a moment when the rockets are not lined up directly opposite each other. At such moments the occupants of any rocket can see neither the serial number nor the clock reading of any other rockets, and photographs cannot be taken.

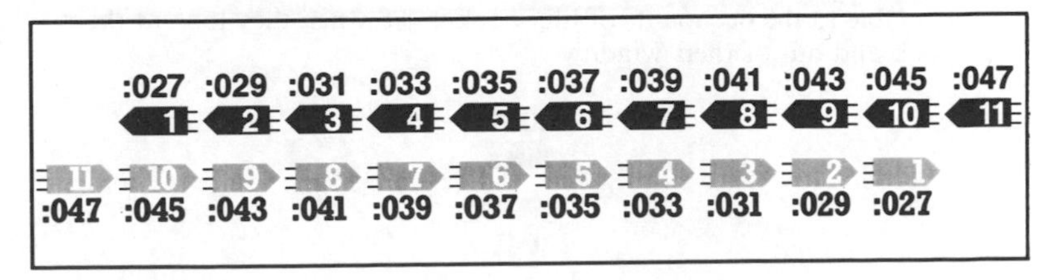

Fig. 5. The two trains, shown at another moment when the occupants of the rockets can learn something about the other train by looking out of their window.

from the data summarized in Fig. 6 that at "red time" :031 blue rocket #8 is directly opposite and its clock reads :041. This statement is based on exactly the same data as the statement in the preceding paragraph, but there is an important difference in how the data are interpreted. The people in the red rocket use their clock reading to establish what the time is (since they believe their clocks to tell correct time) and record the reading of the blue clock as merely a number displayed on a certain timepiece whose relevance to correct time remains to be established. For people on the red train red clocks tell time; blue clocks are objects to be investigated. For people on the blue train, it is the other way around.

Let us suppose, then, that the people in each rocket have cameras that can take such pictures (or sketch pads in which they draw them) as a convenient way of recording the information acquired when a rocket of the other train is directly opposite. *You too have been provided with such a camera, to aid in your investigation of the data collected by the train people.*

Press the left- or right-arrow key, and notice the vertical double arrow (↕) that appears between the two trains (Fig. 7(a)). We shall refer to it as the "camera." If you hold down the left- (or right-) arrow key the camera moves to the right or the left. Should it pass an interesting place release the key and it will stop there.

To take a picture press the F9 key (hereafter referred to as the

Fig. 6. A pictorial summary of the information available either to the occupants of rocket #3 of the red (upper) train or to the occupants of rocket #8 of the blue (lower) train at the moment shown in Fig. 5. Occupants of red rocket #3, while noting that their own clock reads :031, can look out the window of their rocket, see the serial number (#8) of a blue rocket, and look into the window of the blue rocket and see that its clock reads :041. The same information is, of course, available to the occupants of blue rocket #8 when they look at their clock and out of their window.

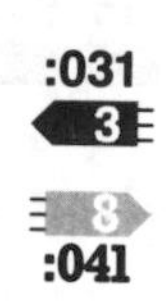

"shutter"). Should the trains not be lined up, you will be reprimanded by a message at the bottom of the screen reminding you that rockets can only take pictures of each other when they are aligned. To make the camera work, let time run forward (or backward) (with the up- or down-arrow keys) until you find an interesting configuration in which the rockets are aligned. Then move the camera (with the right- and left-arrow keys) until it is anywhere between the two rockets whose numbers and clock readings you wish to record. Now, when you press the shutter, a picture representing the data collected by people in either of the rockets you have indicated appears at the lower left of the screen (Fig. 7(b)). This picture contains precisely the same information as the pictures that people in either of the two rockets might have made.

As it happens one cannot learn much of interest from a single picture. But *pairs* of pictures, as we shall see, can be very informative indeed. So take another picture: Search around in time (up- or down-arrows) until you find another (or the same)

Fig. 7. A camera, represented by a double arrow (↕), appears on the screen whenever the right- or left-arrow key is touched, and can be moved right or left by pushing down the appropriate key. When the camera is between a pair of rockets *and the rockets are perfectly aligned* then a picture can be taken by pressing the shutter key (F9). The camera is shown at a configuration of such perfect alignment in (a), and the resulting picture is shown in (b).

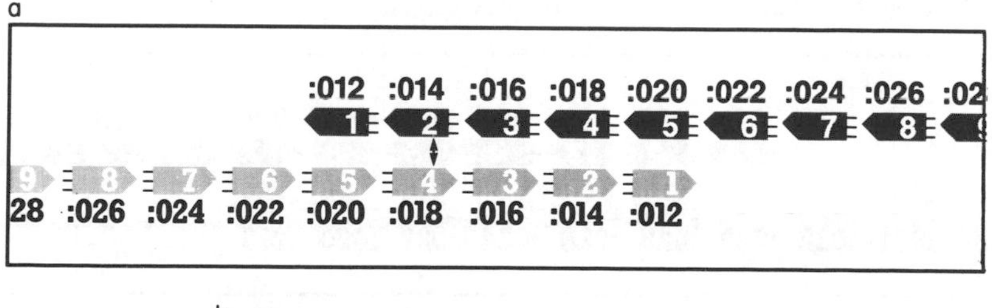

b
:014
2
4
:018

lined up configuration, move the camera right and left (right- or left-arrows) until you find another enticing pair of lined up rockets, press the shutter (F9), and notice that the new picture you have just taken appears at the lower right (Fig. 8).

You can continue taking further pictures in this way, but only two of them will be displayed on the screen at a time. If you take a third picture, the first two will be erased, and the third (and a subsequent fourth) will be displayed in the same parts of the screen where the first and second were shown. It is, therefore, a good idea to note the information contained in a pair of pictures, before going on to take another pair.

The reason only two pictures are shown at a time is partly to avoid cluttering up the screen but, more importantly, because *all* interesting information can be gathered by considering appropriate *pairs* of pictures. It is never necessary to have more than two in front of you at any one time.

E. Drawing conclusions from pictures

Imagine now that the trains are loaded with at least one passenger in each rocket, and that they pass by each other, as pictured on the

Fig. 8. After the picture in Fig. 7(b) was taken, time was advanced (by holding down the up-arrow key), another moment of perfect alignment was found, the camera was moved (by pressing the right-arrow key) to another pair of rockets, and a second photograph was taken (by pressing F9). Both pictures are shown in the figure in the actual positions they occupy on the display screen.

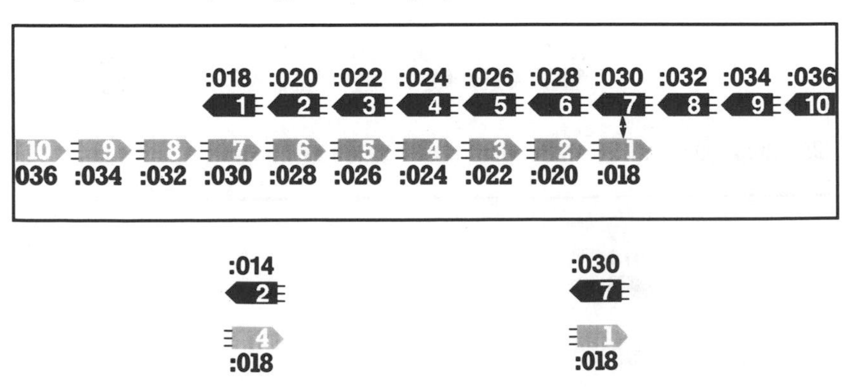

display screen, with passengers in each rocket photographing and noting down what they see as their window passes directly opposite each rocket window of the other train. After the trains pass each other, they return to their base, and all the passengers disembark with their pictures (without being allowed to examine the clocks in any of the other rockets of either train). They are immediately rushed off to two debriefing centers – one for the red passengers and another for the blue, where they can pool all their photographs, and use them to answer a series of questions about the nature and behavior of the other train.

Both groups of passengers have the same collection of photographs at their disposal. But their answers to certain questions can differ, because the red passengers will extract information from the pictures under the assumption that the red clocks were synchronized, while the blue passengers will assume that the blue clocks agreed with each other. You can use the Relativity Engine to answer questions put to the passengers. For example:

Question: How fast do the red passengers say the blue train was moving?

Answer: To answer this you need to follow a particular piece of the blue train as it moves along. Any rocket will do – say rocket #7. Take any two pictures that show rocket #7 at two different moments. Such a pair is shown in Fig. 9. From these two pictures, the red passengers would reason as follows:

Blue rocket #7 was opposite red rocket #2 at red time :023, and a little later, at red time : 83 it was opposite red rocket #14. It therefore moved 12 rocket lengths in a time of 60 ticks, so its speed

Fig. 9. A pair of photographs, from which occupants of the red (upper) train will conclude that rocket #7 of the blue (lower) train is moving at a speed of 0.2 red rockets per tick, and that the clock on blue rocket #7 is running at only 60% the rate of the clocks on the red train.

:023
2
7
:033

:083
14
7
:069

is 12/60 or 0.2 red rocket lengths per tick.[4] *(Problem 1: Find other pictures of blue rocket #7 at other moments that confirm the conclusion that its speed is 0.2 red rocket lengths per tick. Find pictures of other blue rockets that demonstrate to the red passengers that those blue rockets are also moving at a speed of 0.2 rocket lengths per tick. Find still other pairs of pictures that demonstrate to the blue passengers that various red rockets are all moving past them at 0.2 blue rockets per tick.)*

But this is not the only information about the blue train that the red passengers can extract from the two pictures of Fig. 9.

Question: How fast are the clocks on the blue rocket running? (We, of course, know that the clocks on both trains are running at the same rate. But because the clocks on neither train are correctly synchronized, it turns out that occupants of each train reach an interesting conclusion about the rate at which the clocks on the other train are running.)

Answer: Occupants of the red train can study the rate at which the blue clock in blue rocket #7 is running by comparing it with their own clocks in the two pictures. Of course the two pictures show two *different* red clocks – one in rocket #2, the other in rocket #14. The red passengers, however, believe that their clocks are all properly synchronized, and therefore that one clock is as good as another for telling correct time. They, therefore, note from the red clocks that 60 ticks (:083 – :023) elapsed, during which the blue clock in blue rocket #7 only changed its reading by 36 ticks (:069 – :033). They conclude that the blue clock in blue rocket #7 is running at a rate that is only 36/60 or 0.6 of the rate of their own clocks. *(Problem 2: Find other pairs of pictures that permit you to reach the same conclusion about the clock in car #7 and various other clocks on the blue rocket. Find pairs of pictures that lead people on the blue train to conclude that clocks on the red train are running slowly.)*

A somewhat more subtle question is this:

Question: How long do the red passengers say the blue rockets are?

Answer: You can answer this from any pair of pictures in which the red clocks read the *same* time. Consider, for example, the pair of pictures shown in Fig. 10. These reveal (1) that at a red time of :056, the center of blue rocket #3 was directly opposite the center of red rocket #11 and (2) that at the *same* red time of :056 ticks, the center of blue rocket #13 was directly opposite the center of red rocket #5. Consequently, at red time :056 ten full blue rockets (half of #3, half of #13, and all of the nine rockets from #4 to #12) stretched across the same space as only six full red rockets (half of #11, half of #5, and all of the rockets from #10 to #6). But if a line of ten blue rockets stretched across the same distance as a line of six red rockets, then the blue rockets must be only 6/10 or 0.6 the length of the red rockets.

Please note two points, the first profoundly important, and the second a technical detail about using and interpreting the Engine:

(1) The fact that both pictures were taken at *the same* time, according to the red clocks, is absolutely essential in permitting the occupants of the red train to conclude that the blue rockets are shorter than the red rockets. If they did not believe that the two pictures in Fig. 10 were taken at the same time, then they would realize that the blue rockets had moved between the two pictures, and they then could not draw conclusions about the length of the segment of blue train from the middle of rocket #3 to the middle of rocket #13 without correcting for this motion.

Fig. 10. A pair of photographs, from which occupants of the red (upper) train will conclude that the blue (lower) rockets are only 60% as long as the red rockets, and that the clocks on the blue train are out of synchronization by 3.2 ticks per rocket. Note that all the actual pictures show are the serial numbers of the rockets and the readings of their clocks; they do *not* show the ends of the rockets. (More accurate but less beautiful photographs ("unretouched") can be taken by using F5 rather than F9 as the shutter key.) The lengths of the blue rockets are *inferred* from noting that it takes 10 of them to fill up the space occupied by only 6 red rockets.

(2) What about the fact that Fig. 10 shows two rockets, one red and one blue, that are clearly the same length? That, dear reader, is an artifact of the way in which the Engine presents the data for your subsequent use. All a camera actually photographs are the serial numbers and clocks attached to the middle of the two rockets. Both ends of the rockets are too far away to appear in the pictures. Because a list of four numbers is rather abstract, and to make it evident that those numbers are associated with rockets moving in definite directions, the Engine has simply added to each photograph in as nonprejudicial a way as possible (i.e., symmetrically) a schematic representation of the nose cones and exhaust fumes. If this offends or confuses you, you can take more literal pictures using F5 as the shutter key rather than F9. The two pictures of Fig. 10, taken in the unretouched form given by the F5 shutter key are shown in Fig. 11. Evidently they are less obviously associated with rockets, but they do make it quite clear that you cannot learn from a single photograph anything about the comparative lengths of the two rockets. *(Problem 3: Find other pairs of pictures showing moments of time that are simultaneous according to the red clocks, and confirm that they all lead to the conclusion that the blue rockets are 0.6 the length of the red ones. Find pairs of pictures showing moments simultaneous according to the blue clocks, and confirm that those pictures lead to the contrary conclusion that the red rockets are 0.6 the length of the blue ones.)*

Fig. 11. The "unretouched" version of the photographs shown in Fig. 10, to emphasize that a single photograph does not show enough of a rocket to permit a direct determination of its length. The picture on the left, for example, contains no information beyond the fact that at the instant the middle of red rocket #11 was directly opposite the middle of red rocket #3, the clock in the middle of red rocket #11 read :056 ticks and the clock in the middle of blue rocket #3 read :040 ticks. The pictures in Fig. 10 also contain no more information than this, but the full rockets have nevertheless been drawn, simply to make them prettier.

:056
11

3
:040

:056
5

13
:072

The two photographs in Fig. 10 also make it evident to occupants of the red train that the clocks on the blue train are not synchronized, for the pictures portray two blue clocks that read :040 and :072 ticks at a single moment of time (:056) according to the red clocks.

Question: *We* know that the clocks on the blue train are out of synchronization by two ticks per blue rocket. How do the occupants of the red train, thinking that their own red clocks *are* synchronized, characterize the lack of synchronization they observe on the blue train (in ticks per blue rocket)?

Answer: They note that the two blue clocks in Fig. 10 disagree by 32 ticks (:072 – :040) and are ten rockets apart (one is on rocket #13 and the other on rocket #3), so the asynchronization is 32/10 = 3.2 ticks per blue rocket. *(Problem 4: Find other pairs of pictures leading occupants of the red train to the same conclusion and still other pairs that lead occupants of the blue train to conclude that clocks on the red train are out of synchronization by 3.2 ticks per red rocket.)*

F. A meteor passes by

One other thing occasionally relieves the boredom of the passengers. Every now and then a meteor moves along the lengths of both trains, and if a photograph is taken at one of those rare moments when it happens to be moving past a pair of windows just as the windows are opposite one another, then the meteor actually appears in the photograph. By comparing pairs of such photographs it is possible to determine the speed of one and the same meteor, according to the occupants of either (or both) of the two trains. You can produce such meteors for subsequent investigation as follows:

First move the rockets and the camera about until you have the camera (which is now functioning as a site for the creation of a meteor) at a part of the picture where you would like the meteor to appear. Then press F7. You will be asked to specify how many "spaces" the meteor (called an "object") moves in each "time step." (A space is just the amount of screen equal to $\frac{1}{6}$ of a rocket length;

a time step is the interval between one picture and the one produced from it by a single touch of the up-arrow.) This has no obvious relation to the speeds people on either train will assign to the meteor, but it gives you a way to control how fast the meteor moves. If you want a fairly fast meteor, when asked how many spaces per time step type, for example, "5/1" and press the carriage return. The meteor will appear wherever you left the camera, and as time moves forward and backward it will zip back and forth between the trains at a brisk pace.

To get a slow meteor you can enter something like "1/3." Note that now as you advance time you only see the meteor in every third picture. This is a limitation of the display screen (more precisely, the monochrome display screen, but the program written for the graphics display does not take advantage of its more sophisticated capabilities), which cannot move the meteor by a fraction of a space. Consequently, if its speed is a third of a space per time step, all the screen can show is that every three time steps the meteor has moved over another space.

The game is now to try to put down meteors in suitable places and with suitable speeds so they can be caught in pairs of photographs. (Once you have such a pair, you can use them to determine the speed assigned to the meteor by occupants of either train.) Some experimentation is required to produce such a meteor. Sometimes you will find that your meteors never appear opposite rockets' windows when the windows themselves are opposite each other, and such meteors cannot be photographed at all. But by adjusting the position of the pointer-camera at the moment you create the meteor, you can make meteors that do appear at windows when photography is possible. (To get rid of an unsatisfactory meteor press F8, or, if you have immediate plans for introducing another one, just press F7 to override the earlier specifications.) A pair of such photographs of a meteor moving at one space per three time steps ($\frac{1}{3}$ of a space per time step) is shown in Fig. 12.

There are many things you can investigate with suitably chosen meteors, of which only two are mentioned here:

(1) It is possible to specify a special meteor speed that results in occupants of *both* the blue and the red trains, using their own clocks, assigning *exactly the same speed* to such meteors. You should determine, by experimentation, what that speed is, in spaces per time step, and you should also note the speed, in rockets per tick, that occupants of either train assign to meteors moving with that special speed.

(2) Interesting things happen if you introduce meteors with speed significantly greater than the special speed. You should try, for example, to produce two photographs of such a meteor in which the time order in which the photographs were taken depends on whether one believes the red or the blue clocks agree with each other. Evidently if the meteor were a real meteor (which was burning up) it would be evident from its appearance which picture was *really* taken first. Thus, if meteors of sufficiently high speed, which were changing in a manner that revealed the direction of time, could be photographed by both trains, this would provide enough information to convince the occupants of one of the trains that there was something wrong with their clocks. (Such meteors, traveling in the opposite direction, would provide similar information to the occupants of the other train.)

(3) Convince yourself, by trying out several cases, that the above problem never arises with meteors moving slower than the special speed.

(Problems: Find pairs of pictures illustrating these points. Can

Fig. 12. A pair of pictures capturing two moments in the history of a meteor (the squarish blob) moving at one space every three time steps. Evidently people on the red (upper) train believe that the meteor moves four rockets in $45 - 19 = 26$ ticks, or at a speed of 2/13 rockets per tick. People on the blue (lower) train say its speed is 1/11 rockets per tick.

you discover (this will be hard, without advice from informed people) a general formula relating the two speeds occupants of the two trains assign to one and the same meteor, that is valid for all the meteors you can create?)

G. Summary of input to the engine

I list here all the ways in which you can communicate with the Engine:

Up-arrow: One step forward in time. (One "time step.")

Down-arrow: One step backward in time.

Right-arrow: Call up the camera and move it one space to the right.

Left-arrow: Call up the camera and move it one space to the left.

F1: Exit from the program.

F5: Take an unretouched photograph (in which the front and rear of the rockets are not in the picture).

F9: Take a retouched photograph (in which the front and rear of the rockets have been added by an artist, to make the direction of motion of each rocket evident from the picture).

F6: Erase any photographs currently on the screen.

F7: Create a meteor at the position of the camera (↕). The camera should be put in the desired position using the right- and left-arrow keys, before F7 is pressed. To get a meteor moving a certain number of spaces in some other number of time steps you must type the number of spaces (preceded by a minus sign if you want a meteor going to the left), then a slash (/), then the number of time steps, and then a carriage return.

F8: Remove the meteor.

IV. Further remarks for sophisticates

A. Pedagogical matters

The above User's Manual (Sec. III) is not intended to be a complete list of all the measurements and discoveries students can make with the Engine, but it should give an idea of the kinds of questions they can be led to explore with its aid. There are two distinct pedagogical strategies for employing the Engine:

(1) Use it as a prologue to a discussion of relativity. Introduce the Engine from an entirely Newtonian point of view as a way of exploring the kinds of false conclusions people can be led to if they fail to realize that their clocks are not synchronized. Extract length contraction, time dilation, and the invariant velocity. Extract, if you want, the relativistic addition law. Get your students used to using the Engine to measure the lengths of moving objects with unsynchronized clocks. Then (and only then) introduce the real world through the constancy of the velocity of light. Point out the lesson of the Engine: that this would not cause problems if different observers used differently synchronized clocks. Prior experience with the Engine will go a long way toward alleviating that feeling of being led into a maze of contradictory assertions that can so easily paralyze even the brightest students on a first exposure to the subject.

(2) Use the Engine as an antidote to such paralysis after the subject has been developed from a conventional point of view. This has the disadvantage of failing to take full advantage of the Engine's therapeutic powers, but the advantage of enabling more capable students to approach the Engine with the relativistic formulas for length contraction, time dilation, relativity of simultaneity, and velocity addition, already at hand, thereby permitting them to undertake their *gedanken* experimental studies with the Engine with the aim of confirming specific quantitative expectations.

It would be futile for me to recommend either course of pedagogical action. The point of this essay is simply to call attention to the Engine as a novel and (in my experience) useful educational device for teaching special relativity. People using it to teach a class will want to play with it themselves, develop their own sets of questions to be explored, and fit it into their syllabus as best meets their own aims.

B. Alternative Engines

The particular version of the Engine described here (and embodied in my program) has the trains moving with respect to each other at the single speed of $\frac{4}{5}c$. One can design more sophisticated Engines with adjustable relative speeds, for screens with greater resolution than a PC (and processors with greater speeds), although not many speeds are conveniently adapted to some of the constraints imposed by the discreteness of any visual display.

I am not persuaded that designing an Engine with variable speed is worth the effort. The most important pedagogical purpose served by the Engine is forcibly bringing home the consequences of making measurements with improperly synchronized clocks. Any single speed can serve this purpose. The other advantage of the present Engine is its widespread accessibility. Any student who uses a PC compatible machine to write essays and term papers can use that same machine to run the Engine. Should the age dawn (as no doubt it will) when students routinely prepare term papers on their own micro-Vaxes, then the time will perhaps have arrived to write a more flexible, powerful, and certainly more colorful Relativity Engine.

Nevertheless, should anybody wish to design a Relativity Engine with parameters different from mine, the following considerations might be of some help:

Regard a computer screen as composed of horizontal rows of discrete cells. The standard IBM monochromatic display, for which my version of the Engine was composed, contains 80 cells per row. For maximum verisimilitude, any version of the Engine should have each train shift by just one cell in each time step. There is also no reason not to choose the time scale so that each clock advances by one unit (a "tick") in each time step. The free parameters in any realization of the Engine are therefore only two:

(1) The number of cells N making up a rocket. If every pair of red and blue rockets is to have a moment of perfect alignment for picture taking, it is necessary that N be even.

(2) The number of ticks Δ by which clocks on adjacent rockets disagree. If any clock is to be capable of showing the same readings as any other clock, then Δ must be an integer. If one is willing to use an Engine with an unpleasant artificial asymmetry in what different clocks can show, then Δ can be merely rational, but it had better have a fairly small denominator, if pairs of pictures taken at the same train time are to be readily available.

The two numbers N and Δ completely determine all the other parameters that characterize the Engine (most importantly, the value of the invariant velocity c and the value of the speed v of one train with respect to the other). If one imposes the additional condition that an object moving at the invariant velocity should be displayed on the screen at every time step – i.e., that such an object should move a whole number of cells in a single time step – then the freedom in choosing the parameters is further reduced.

To work this out, consider the progress of the two pairs of rockets depicted in Fig. 13. Between Figs. 13(a) and 13(b) $\frac{1}{2}N$ time steps have elapsed (as indicated by the clock readings adjoining the rockets). Inspecting these two parts of the figure reveals that the speed of either train with respect to the other is

$$v = 1/(\Delta + \tfrac{1}{2}N) \tag{1}$$

rockets per tick. Further inspection reveals that in a time $\Delta + \frac{1}{2}N$ according to one train, a given clock on the other has only advanced by $\frac{1}{2}N$ ticks and, therefore,

$$\sqrt{1 - v^2/c^2} = \tfrac{1}{2}N/(\tfrac{1}{2}N + \Delta). \tag{2}$$

Equations (1) and (2) together give

$$c^2 = 1/[\Delta(N + \Delta)] \tag{3}$$

rockets per tick. An easy way to extract the condition that an object moving at speed c should cover a whole number of cells per time step is to note that the speed u of either train in the *screen* ("platform") frame is just one cell per time step so we require c/u in cells per time step to be an integer n. Now Eq. (3) gives us c in rockets per tick. We can get u in the same units, by noting that u is also just the speed of an object fixed on the screen in either *train* frame. It follows from Figs. 13 (a) and 13 (c) that in rockets per tick,

$$u = 1/(N + \Delta). \tag{4}$$

Combining this with Eq. (3) we have

$$n^2 = c^2/u^2 = 1 + N/\Delta. \tag{5}$$

The general Engine can therefore be characterized as follows.

Given the time shift Δ between adjacent rockets, the length of each rocket should be taken to be

$$N = \Delta(n^2 - 1), \tag{6}$$

Fig. 13. A more abstract representation of three moments in the history of a pair of rockets from each train. Each rocket moves one cell per time step and stretches over N cells. Consequently $\frac{1}{2}N$ time steps elapse between parts (a) and (b) and between parts (b) and (c). This is confirmed by the fact that each clock advances by $\frac{1}{2}N$ ticks between parts (a) and (b) and between parts (b) and (c). Note also that the clocks on each train are out of synchronization by Δ ticks per rocket.

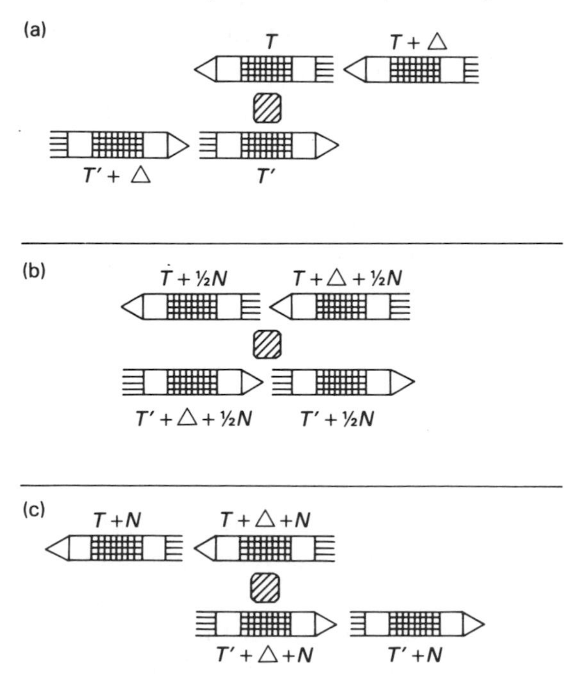

where n is any integer greater than 1, and Δ (which is integral in the most attractive Engines) must be such as to make N an even integer. The invariant velocity is then

$$c = 1/n\Delta \tag{7}$$

rockets per tick, and the relative speed v of the two trains is given by

$$v/c = 2n/(n^2 + 1), \tag{8}$$

resulting in

$$\sqrt{1 - v^2/c^2} = (n^2 - 1)/(n^2 + 1). \tag{9}$$

My version of the Engine is one of the simplest possible, with $n = \Delta = 2$.

C. Computational details

The particular realization of the Engine I constructed runs on the IBM PC. There are two versions, designed for either the ordinary monochrome display or the color graphics display. Color helps to distinguish the different frames of reference and is much prettier to watch,[5] but monochrome is more commonly available and quite manageable, thanks to the IBM extended character set and the graphical possibilities of inverse video.

The program is written in ordinary PC BASIC, which has to be compiled if it is to run at anything like a reasonable speed. I would be happy to send anybody a ready-to-run copy, upon receiving a formatted $5\frac{1}{4}$-in. floppy disk and a suitable container, stamped and addressed. In return for providing you with your own Engine, I would appreciate receiving any suggestions you have for its improvement.

People interested in writing their own Engines for the PC might note that to achieve a decent speed I also found it necessary to execute the graphics by poking the appropriate data directly into the memory locations where the display screen is mapped. As an intelligible source of the arcane information necessary to achieve this feat, I found Peter Norton's *Inside the IBM PC* quite satisfactory.

Notes and references

1 In this respect, the Engine resembles the Brehme diagrams. It operates, however, at a very much more intuitive level, by keeping time as time rather than representing it as a second spatial direction.

2 A very early asymmetric attempt at a Relativity Engine is described in N. D. Mermin, *Space and Time in Special Relativity* (McGraw-Hill, New York, 1968, reissued by Waveland Press, Prospect Heights, Illinois, 1989), Chap. 12.

3 A word to sophisticates: These rockets are moving with uniform velocity, so why are their engines on? Primarily because it makes the pictures prettier and more vivid. If this bothers you, then please regard the rockets as moving not through the "void," but through a resistive medium. The reason they are not accelerating in spite of the blast is that the thrust of the rockets just balances the frictional force.

4 The passengers on the red train do not, of course, use the term "red time," but simply say "time," since for them time is what they read from their clocks, which they believe to be synchronized. We, being better informed, may find it useful to use more accurate terms like "red time" or "blue time" to remind us that such statements actually refer only to the readings of a particular set of clocks.

5 Something with the resolution of the IBM Enhanced Color Display is required to make the Engine easily legible; the more primitive IBM color display also produces some unattractive noise between time steps.

19

Relativistic addition of velocities directly from the constancy of the velocity of light

Greenwood[1] has given a derivation of the relativistic addition law for parallel velocities that makes no explicit use of the Lorentz transformation equations, relying instead on direct applications of length contraction and time dilation. This kind of approach is necessary in a general education physics course, where more abstract derivations can only obscure the physics. Arguing from the Lorentz transformation in such courses is as inappropriate as trying to teach school children elementary geometrical facts about circles and triangles by starting from their algebraic representations in Cartesian coordinates.

In this spirit I would like to describe another derivation of the addition law that dispenses not only with the Lorentz transformation, but also makes no use of length contraction, time dilation, or the relativity of simultaneity. This argument[2] extracts the result as an immediate consequence of the constancy of the velocity of light.

As in Greenwood's *gedanken* experiment, we consider a sequence of events taking place on a long straight uniformly moving train. I shall describe all the events from the viewpoint of a frame of reference S in which the train moves parallel to its own length with speed v. All references to speeds, distances, lengths, and times refer to their values in the frame S.

At a certain instant a photon (speed c) and a massive particle (speed w less than c, but greater than v) begin a race from the rear of the train to the front. The photon, of course, wins, taking a time

T to reach the front. There, however, it is immediately reflected back toward the rear, and a time T' after the reflection it encounters the particle, which is still making its way toward the front (see Fig. 1).

The place on the train where the two meet is behind the front end by a fraction f of the full length of the train. Note that f is a frame-independent invariant, since there can be no dispute about where on the train the meeting occurred. We can calculate this invariant in terms of c, w, and v from three simple facts.

(a) The total distance the particle moved from the start of the race to its encounter with the photon is equal to the distance the photon moved in going from the rear to the front minus the distance the photon moved in going from the front back to the particle:

$$w(T+T')=c(T-T'). \tag{1}$$

(b) The distance the photon moved in going from the rear to the front is just the length L of the moving train augmented by the distance the train moved while the photon was so moving:

$$cT=L+vT. \tag{2}$$

Fig. 1. Three moments of time in frame S. T is the time between start and reflection. T' is the time between reflection and reencounter. f is the fraction of train between the place of reencounter and the front of the train. In this case $f=\frac{2}{3}$.

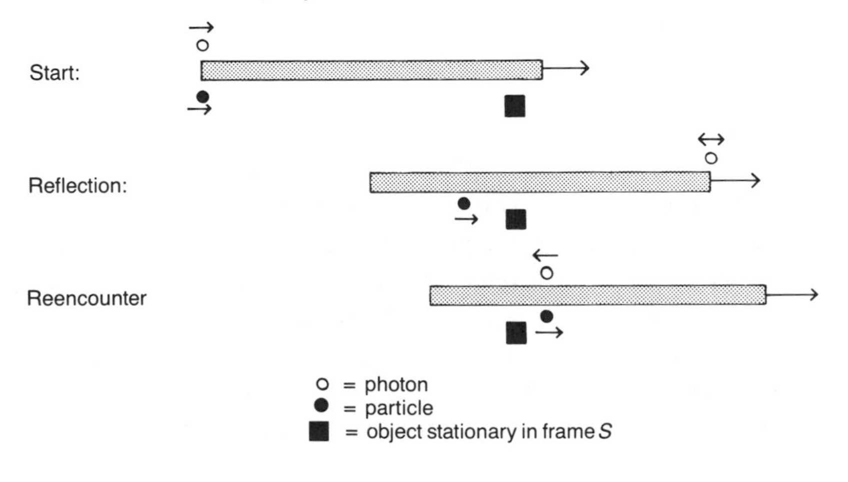

(c) The distance the photon moved in going from the front to the particle is just the length of train from the front to the meeting point on the train, diminished by the distance the train moved while the photon was so moving:

$$cT' = fL - vT'. \tag{3}$$

Equation (1) immediately gives $T'/T = (c-w)/(c+w)$. On the other hand, eliminating the unknown L between (2) and (3) gives $T'/T = f(c-v)/(c+v)$. Since these two expressions for T'/T must agree we conclude that[3]

$$f = \frac{(c+v)(c-w)}{(c-v)(c+w)}. \tag{4}$$

Nowhere in the derivation of (4) did we use the fact that the velocity v of the train was nonzero. The result is therefore equally valid in the frame of reference S' that moves with the train. Calling the speed of the particle in the train frame u, noting that c and f are invariants, and evaluating the result (4) in the train frame where $v = 0$, we have[4]

$$f = \frac{(c-u)}{(c+u)}. \tag{5}$$

Setting equal the two expressions (4) and (5) for the invariant f immediately gives the addition law in its multiplicative form,

$$\frac{c+w}{c-w} = \left(\frac{c+u}{c-u}\right)\left(\frac{c+v}{c-v}\right), \tag{6}$$

or, solving (6) for w,

$$w = \frac{u+v}{1+uv/c^2}. \tag{7}$$

A case can be made for presenting this argument to a naive class even before deriving any of the other relativistic effects. By doing so one immediately establishes that all speeds – not just that of light – must transform unexpectedly under a change of frame, while at the same time revealing the reassuring continuity with the results of nonrelativistic intuition as the speeds diminish. The argument also gives an initial familiarity with precisely the kind of analysis from which the basic facts about clock synchronization and simultaneity can immediately be extracted, making it easier to concentrate on the genuine conceptual subtleties without losing them in the confusion engendered by the elementary algebra.

Notes and references

1 M. S. Greenwood, *Am. J. Phys.* **50**, 1156 (1982).

2 The argument can be found, in unhelpful brevity, as homework problem 5 in N. D. Mermin, *Space and Time in Special Relativity* (McGraw-Hill, New York, 1968, reissued by Waveland Press, Prospect Heights, Illinois, 1989), p. 230. Although I have never encountered this formulation anywhere else, it must surely have made earlier appearances.

3 The confidence of students in (4) can be enhanced by noting that it gives the expected answer when $w = v$ (the particle never leaves the rear of the train) and when $w = c$ (the race is a tie). It is worth emphasizing to students that nothing "peculiar" whatever goes into the derivation of (4) – Galileo would have reached the same conclusion. The only part of the entire argument containing a post-1905 thought is the assumption that (4) can be used with the *same* value for c in *any* frame moving parallel to the train. More enterprising students might be invited to generalize (4) to the case where the "photon" has speeds c_F toward the front and c_R toward the rear, exploring how the nonrelativistic addition law can be recovered when c_F and c_R are not taken to be invariant.

4 The grandeur of this derivation of (5) from (4) is surely an important part of any student's general education, but it may be too breathtaking the first time through. If so, the instructor can simply repeat steps (a) – (c) in the train frame (where the algebra is even simpler), emphasizing only afterwards the shorter and more elegant route.

20

Relativity without light

I. Introduction

The first statement of the special theory of relativity[1] was intimately tied to the first unambiguous assertion that the velocity of light in empty space has the value c, independent of frame of reference. Reflecting this historical circumstance, light has almost always had a central role in subsequent expositions of relativity, playing, for example, an important part in establishing a convention for the synchronization of distant clocks, or determining the rate of moving clocks or the length of moving rigid rods.

Relativity, however, is not a branch of electromagnetism and the subject can be developed without any reference whatever to light. Since this is not the conventional approach, I should emphasize that in asserting that relativity can be developed without light, I do not have in mind the trivial sense in which anything else that moves at the invariant velocity c can serve as well. Nor do I mean merely that the theory can be built out of the fact that the relation of lightlike separation between a pair of events is invariant under change of frame, regardless of whether or not there exists any form of matter or energy that actually can propagate at the invariant velocity.

What I do mean, and shall show below, is that a parallel velocity addition law of the form

$$w = (u + v)/(1 + Kuv), \qquad (1.1)$$

with K a universal non-negative constant, is the most general possible compatible with the principle of relativity, supplemented only by certain natural assumptions of homogeneity, isotropy, and smoothness.[2] It follows, of course, from (1.1) that $K^{-1/2}$ is an invariant velocity. Thus Einstein's second postulate is a consequence of his first,[3] if it is stated generally in terms of an invariant velocity rather than specifically in terms of the behavior of light.[4]

From this point of view, experiments establishing the constancy of the velocity of light are only significant because they determine the numerical value of the parameter K. Because that value turned out to be the inverse square of the speed of light in empty space, rather than the expected Galilean value $K=0$, the impact of such experiments was, of course, revolutionary, and light became an essential part of the new theory of space and time. However, the value of K can be determined from careful measurements of the speed of any moving object from two inertial frames in relative motion.[5] Measurements on light offer an especially elegant and precise route to determining the parameter K appearing in the addition law (1.1), but the law itself, as I shall show, follows from the principle of relativity and the fundamental relations between distance, time, and velocity, without the need of any additional facts or postulates.

There are pedagogical as well as conceptual advantages to eliminating light from its central role in relativity theory. By starting only with the principle of relativity and refraining from trying to relate temporal judgments made in different frames of reference, one is led directly to (1.1) which lies, of course, at the heart of special relativity, without ever having to face the distracting sense of paradox that bedevils more conventional attempts from the very first steps. The price is a somewhat higher level of analysis: a little elementary calculus is unavoidable. The approach is therefore unavailable for a general education physics course, but as an introduction to special relativity for physics majors I believe it has much to recommend it.

In what follows I shall use without critical analysis the concept of an inertial frame of reference and, within any given inertial frame, the concepts of distance, time, and velocity. I shall use these last three notions only through the relation that the distance covered by a uniformly moving object in a given time is the product of its velocity with the time, a connection that any more rigorous development must surely preserve. What logical rigor I bring to the argument will be entirely negative: I shall scrupulously avoid making any assumptions of how distances, times, and velocities in one inertial frame are related to those in another. It is possible to be more systematic and more economical in basic concepts, but to do so here would distract from the central point that the existence of an invariant velocity is not required as an independent assumption.

My derivation of the addition law (1.1) from the principle of relativity blends *gedanken* experiment and analysis. The *gedanken* experiments are of the usual sort, except that no light signals or particles ("photons") moving at special invariant velocities ever appear. The analysis is rather less familiar and more intricate; without the second postulate one has to work harder.

I present in Sec. II that part of the analysis that is of a general character, independent of any *gedanken* experiment, reducing the problem of finding the function of two variables specifying the general addition law to that of finding a function of a single variable. The *gedanken* experiments are designed to determine that unknown function.

I describe two such *gedanken* experiments in Secs. III and IV, each of a somewhat different character and leading by different routes to the same conclusion (1.1). There is, of course, no logical need for two independent arguments. I include the first because it is a direct generalization of an earlier argument[6] that very simply and economically extracts the addition law from first principles *using* the principle of the constancy of the velocity of light. The second argument is, I suspect, a better starting point to give the whole procedure a tighter logical structure, since it deals directly with the relation between periodic processes in different frames, and thus involves in a direct and fundamental way those aspects of phenomena that are encompassed with such deceptive simplicity in the general notion of time.

The arguments in Secs. III and IV are carried to the point where they can be concluded by the single discussion I give in Sec. V.

II. The general form for a velocity addition law

We consider various objects and frames of reference that move uniformly along a single common direction. An addition law is a relation

$$w = f(u, v) \tag{2.1}$$

between the velocity w of an object in frame A, its velocity u in frame B, and the velocity v of frame B in frame A.

We can identify the velocity of the uniformly moving object with that of its proper frame C, and regard (2.1) more systematically as a relation between the relative velocities of various frames of reference, writing it in the form

$$v_{CA} = f(v_{CB}, v_{BA}). \tag{2.2}$$

The principle of relativity is, of course, implicit in the assumption that f depends only on the relative velocities v_{CB} and v_{BA}.

In the absence of any distinction between the two directions of motion, three velocities related by (2.2) must continue to be related if all three signs are changed, so f must be an odd function:

$$f(-x, -y) = -f(x, y). \tag{2.3}$$

The same symmetry also requires that

$$v_{XY} = -v_{YX}. \tag{2.4}$$

In view of the last two relations we find, interchanging A and C in (2.2), that

$$\begin{aligned} f(v_{AB}, v_{BC}) &= v_{AC} = -v_{CA} = -f(v_{CB}, v_{BA}) \\ &= f(-v_{CB}, -v_{BA}) = f(v_{BC}, v_{AB}); \end{aligned} \tag{2.5}$$

i.e., f must be symmetric in its arguments:

$$f(x, y) = f(y, x). \tag{2.6}$$

Another important property of f follows from introducing a fourth frame and noting that v_{DA} can be expressed in two different ways:

$$f(v_{DB}, v_{BA}) = v_{DA} = f(v_{DC}, v_{CA}). \tag{2.7}$$

Expanding v_{DB} on the left and v_{CA} on the right,

$$\begin{aligned} v_{DB} &= f(v_{DC}, v_{CB}), \\ v_{CA} &= f(v_{CB}, v_{BA}), \end{aligned} \tag{2.8}$$

we find the general condition that

$$f(f(x,y),z) = f(x, f(y,z)). \tag{2.9}$$

All of the above relations follow from simple considerations of symmetry. I now also assume that the addition law we seek is expressed by a *smooth* function f. But if f is continuous and differentiable, then we can express it in terms of a function of a single variable.

To do this we first define

$$f_2(x,y) = \frac{\partial f(x,y)}{\partial y}. \tag{2.10}$$

Differentiating (2.9) with respect to z gives.

$$f_2(f(x,y),z) - f_2(x, f(y,z)) f_2(y,z). \tag{2.11}$$

Setting z to zero in (2.11) gives

$$f_2(f(x,y),0) = f_2(x,y) f_2(y,0), \tag{2.12}$$

where we use the fact, evident from the original definition of f, that

$$f(y,0) = y. \tag{2.13}$$

Let us now fix x and regard $f(x,y)$ as a function f of the single variable y, depending parametrically on x. Let us regard $f_2(y,0)$ as a second function of y. The structure of (2.12) is then that of the ordinary differential equation

$$f_2(f,0) = \frac{\mathrm{d}f}{\mathrm{d}y} f_2(y,0), \tag{2.14}$$

or

$$\mathrm{d}y/f_2(y,0) = \mathrm{d}f/f_2(f,0). \tag{2.15}$$

Defining a new function $h(z)$ by

$$h(z) = \int dz/f_2(z,0), \tag{2.16}$$

we see that (2.15) requires that

$$h(f) = h(y) + \text{const}, \tag{2.17}$$

where the constant is independent of y but can depend on the parameter x. The symmetry (2.6) now requires that the constant be precisely $h(x)$ (plus a genuine constant that can be absorbed into a redefinition of h).[7] We conclude that there must be a function h of a single variable such that

$$h(f(x,y)) = h(x) + h(y), \tag{2.18}$$

so that the form of the addition law is

$$f(x,y) = h^{-1}(h(x) + h(y)). \tag{2.19}$$

To determine the addition law it therefore suffices to determine the function h. For this purpose it is enough to know the function $f(x,y)$ in the neighborhood of $y=0$, since (2.16) gives

$$h'(z) = \left.\frac{1}{\partial f(z,y)/\partial y}\right|_{y=0}. \tag{2.20}$$

The condition (2.13) is consistent with (2.18) only if

$$h(0) = 0, \tag{2.21}$$

which provides the boundary condition necessary to determine h from (2.20) by integration.

We turn now to some *gedanken* experiments which enable us to determine the form of $h(z)$ to within a single universal constant K.

III. Precise form of the addition law: a *gedanken* race

We consider a race taking place within a long straight train (Fig. 1). A tortoise and a hare start at the rear of the train toward the front. The hare gets there first, turns immediately around, and, racing back towards the rear, encounters the tortoise still making its way toward the front.[8] Let u be the speed of the tortoise in the train frame and s, the speed (in either direction) of the hare.

The part of the train where the two meet again is behind the front end by some fraction r of the full length of the train. That fraction is a frame-independent invariant, since there can be no disputing where on the train the meeting occurred. (They might, for example, meet in the 73rd car from the front of a train consisting of 100 identical cars, giving the value $r = 0.73$. Only passengers in car 73 would testify to having observed the encounter, and this testimony would be acceptable to observers in any frame of reference, even though they might have quite different ideas about the lengths of the cars.)

Let us now calculate r in a frame (the "v-frame") in which the train moves with speed v (Fig. 2) and examine the consequences of the fact that r cannot depend on v. It will suffice to consider v less than u, so we can take the direction of motion of the hare on its

Fig. 1. The race between the hare (black sphere) and tortoise (white sphere) as described in the train frame. The speed of the hare (in either direction) is s, and the speed of the tortoise is u. The distance between the front (right) end of the train and the final meeting place of hare and tortoise is a fraction r of the total length D of the train.

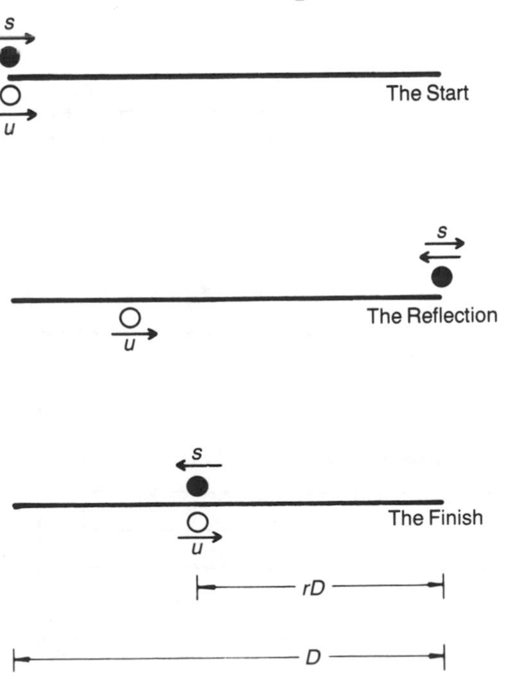

return trip to be opposite to that of the train in the v-frame. Let w be the v-frame speed of the tortoise, and let s_1 and s_2 be the v-frame speeds of the hare on its way toward and back from the front of the train. These speeds are related to the train-frame speeds by the addition law

$$w = f(u, v), \qquad s_1 = f(s, v), \qquad s_2 = f(s, -v). \qquad (3.1)$$

(The last of these follows from noting that the *velocities* of the

Fig. 2. The race between the hare (black sphere) and tortoise (white sphere) as described in the v-frame, in which the train moves to the right with speed v. The speed of the hare is s_1 to the right and s_2 to the left; the speed of the tortoise is w. The length of the train is L. The time between the start and the reflection is T, and the time between the reflection and the finish, T'. Various lengths appearing in Eqs. (3.2)–(3.4) are indicated.

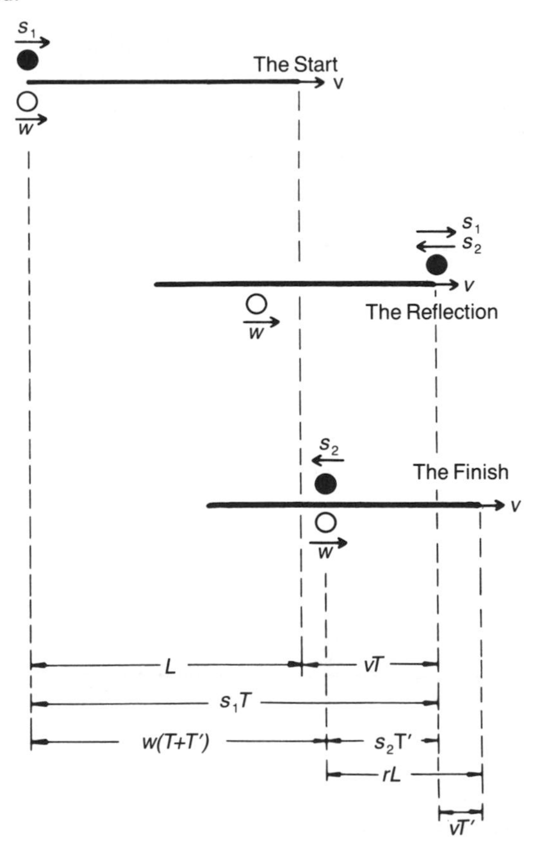

hare in the two frames are $-s$ and $-s_2$, so we have $-s_2 = f(-s, v) = -f(s, -v)$, in view of the oddness of f.)

Let T be the time (in the v-frame) it takes the hare to get from the rear of the train to the front, and T' the time it takes the hare to get from the front back to the tortoise. Let L be the (v-frame) length of the train. We do not know the values of T, T', or L, but it does not matter – they will drop out of the final result. To express the quantity r entirely in terms of the speeds w, s_1, and v, we need only note the following:

(a) The total distance the tortoise covers from the start of the race to its reencounter with the hare is $w(T + T')$. This, however, is equal to the distance the hare covers in going from the rear to the front, $s_1 T$, minus the distance the hare goes back towards the rear before meeting the tortoise, $s_2 T'$:

$$w(T + T') = s_1 T - s_2 T'. \tag{3.2}$$

(b) The distance the hare moves in going from the rear to the front is just the length of the train augmented by the distance the train moves while the hare is so moving:

$$s_1 T = L + vT. \tag{3.3}$$

(c) The distance the hare moves in going from the front back to the tortoise is just the length of the train from the front to the meeting point on the train, rL, diminished by the distance the train moves while the hare is so moving:

$$s_2 T' = rL - vT'. \tag{3.4}$$

We can solve (3.3) and (3.4) for T and T', substitute the resulting expressions in (3.2), cancel the factor L common to all terms, and solve the resulting expression for r to find[9]

$$r = (s_1 - w)(s_2 + v)/(s_2 + w)(s_1 - v). \tag{3.5}$$

Pause to note that the derivation of (3.5) is entirely free of relativistic notions, the entire computation having been done in a single frame (the v-frame). Relativity has informed the argument only through the absence of any attempts to make naive identifications of the v-frame speeds appearing in (3.5) with the train frame speeds s and u, and through the care taken to eliminate from the final expression any v-frame lengths or times.

To find the form of the addition law we must determine the function h appearing in (2.19). A comparison of (3.1) with the expression (2.20) for h' shows that

$$\left.\frac{\partial s_1}{\partial v}\right|_{v=0} = 1/h'(s),$$

$$\left.\frac{\partial s_2}{\partial v}\right|_{v=0} = -1/h'(s),$$

$$\left.\frac{\partial w}{\partial v}\right|_{v=0} = 1/h'(u). \tag{3.6}$$

Also, of course, as $v \to 0$,

$$s_1 \to s, \qquad s_2 \to s, \qquad w \to u. \tag{3.7}$$

Because r is independent of v, $\partial \ln(r)/\partial v$ must vanish. But using (3.6) and (3.7) we can evaluate this quantity from (3.5) at $v=0$:

$$\left.\frac{\partial \ln(r)}{\partial v}\right|_{v=0} = \frac{2su^2}{s^2-u^2}\left[\frac{1}{s^2}\left(\frac{1}{h'(s)}-1\right)-\frac{1}{u^2}\left(\frac{1}{h'(u)}-1\right)\right]. \tag{3.8}$$

For the quantity in (3.8) to vanish it is necessary that

$$(1/s^2)[1-1/h'(s)] = (1/u^2)[1-1/h'(u)]. \tag{3.9}$$

Since the left side of (3.9) depends only on s and the right side only on u, each side must be equal to a constant K independent of s or u, and we conclude that

$$h'(u) = 1/(1-Ku^2). \tag{3.10}$$

Integration of (3.10) leads directly to the addition law (3.1). I defer this to Sec. V, turning in Sec. IV to a somewhat different way of arriving at (3.10).

IV. Precise form of the addition law: a *gedanken* oscillator

Consider a ball rolling back and forth between the rear and front of the train at a fixed (train frame) speed u. We examine this "u-

oscillator" in a frame (the "v-frame") in which the train moves with a speed v less than u. Let $t_1(u, v)$ and $t_2(u, v)$ be the v-frame times for the back-to-front and front-to-back parts of each cycle. We first establish that the difference between these times is independent of the train frame speed u of the ball:

$$t_1(u, v) - t_2(u, v) \text{ independent of } u. \tag{4.1}$$

This is, of course, obvious in the train frame, where the difference is, in fact, identically equal to zero. Since we make no assumptions about how statements of time in different frames are related, however, we must establish (4.1) directly in the v-frame.

Since the time difference in (4.1) must be a continuous function of u, it is enough to establish that it is the same for two values, u and u' whose ratio is the ratio of two odd integers[10]:

$$u = (2m + 1)u_0, \qquad u' = (2n + 1)u_0. \tag{4.2}$$

Let the balls start together at the rear of the train. Since u and u' are train-frame speeds, it is evident in the train frame that after exactly $m + \frac{1}{2}$ complete cycles of the first ball and $n + \frac{1}{2}$ of the second, the two will arrive together at the front of the train; and after another $m + \frac{1}{2}$ cycles of the first and $n + \frac{1}{2}$ of the second, they will again arrive together at the rear. Since these facts can be verified simply by counting round trips and noting the presence of two balls in the same place at the same time, they must be valid in any frame. Hence the v-frame times T_1 and T_1' that it takes the balls to complete their first $m + \frac{1}{2}$ and $n + \frac{1}{2}$ respective cycles must be equal, as must the corresponding times T_2 and T_2' for the next $m + \frac{1}{2}$ or $n + \frac{1}{2}$ cycles. We then certainly have

$$T_1 - T_2 = T_1' - T_2'. \tag{4.3}$$

But T_1 is the v-frame time for m round trips plus a back-to-front trip, while T_2 is the time for m round trips plus a front-to-back trip. The round-trip times drop out of the difference, leaving

$$T_1 - T_2 = t_1(u, v) - t_2(u, v). \tag{4.4}$$

In the same way

$$T_1' - T_2' = t_1(u', v) - t_2(u', v). \tag{4.5}$$

Collectively (4.3)–(4.5) assert that

$$t_1(u,v) - t_2(u,v) = t_1(u',v) - t_2(u',v), \tag{4.6}$$

which is precisely the content of (4.1).

We now express the time difference appearing in (4.1) in terms of the function f appearing in the addition law. Note first that in the v-frame the ball moves from rear to front with speed $f(u,v)$ taking a time $t_1(u,v)$ to cover a distance equal to the (v-frame)

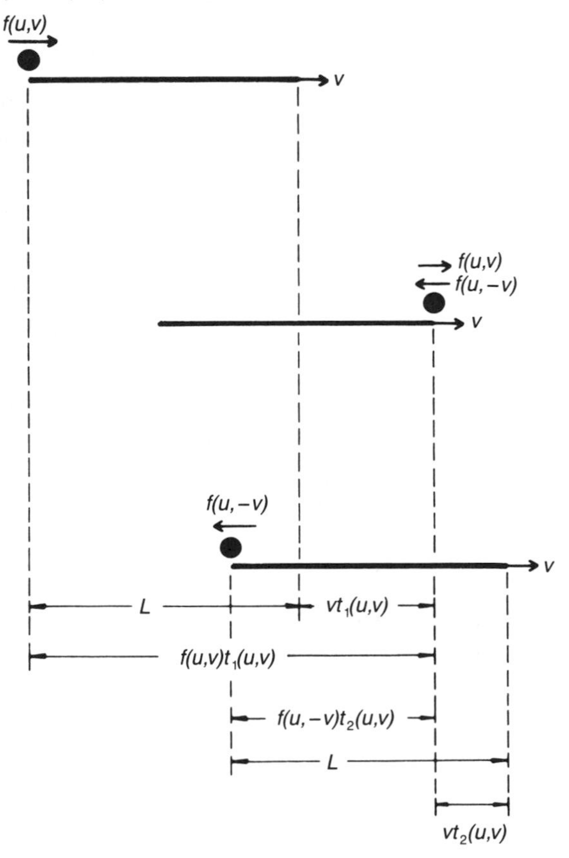

Fig. 3. A *gedanken* oscillator, as described in the v-frame, in which the train moves to the right with speed v. The speed of the ball is $f(u,v)$ to the right and $f(u,-v)$ to the left. The length of the train is L. The time between the top and middle pictures is $t_1(u,v)$; the time between the middle and bottom pictures is $t_2(u,v)$. Various lengths appearing in Eqs. (4.7) and (4.8) are indicated.

length L of the train plus the distance $vt_1(u, v)$ moved by the front of the train in that time (Fig. 3):

$$f(u, v)t_1(u, v) = L + vt_1(u, v). \tag{4.7}$$

The ball next moves from front to rear with speed $f(u, -v)$, taking a time $t_2(u, v)$ to cover a distance equal to the length L of the train minus the distance $vt_2(u, v)$ moved by the rear of the train in that time:

$$f(u, -v)t_2(u, v) = L - vt_2(u, v). \tag{4.8}$$

Solving (4.7) and (4.8) for t_1 and t_2, we have

$$t_1(u, v) - t_2(u, v) = L/[f(u, v) - v] - L/[f(u, -v) + v]. \tag{4.9}$$

But since L, the v-frame length of the train, does not depend on u, the train-frame speed of the balls, we can conclude from (4.9) and (4.1) that

$$1/[f(u, v) - v] - 1/[f(u, -v) + v] = g(v), \tag{4.10}$$

where the function $g(v)$ cannot depend on u.

Dividing by $2v$, we can rewrite (4.10) as

$$\frac{g(v)}{2v} = \frac{\{[f(u, -v) - f(u, v)]/2v\} + 1}{[f(u, v) - v][f(u, -v) + v]}. \tag{4.11}$$

As v approaches zero the left side of (4.11) approaches some value K, and (4.11) gives

$$K = \frac{1 - [\partial f(u, v)/\partial v]|_{v=0}}{u^2}, \tag{4.12}$$

which (cf. (2.20)) again gives us the result

$$h'(u) = 1/(1 - Ku^2). \tag{4.13}$$

V. Relativity without light

Using the boundary condition (2.21) we can integrate (3.10) or (4.13) to get

$$h(z) = \frac{1}{2K^{1/2}} \ln\left(\frac{1 + K^{1/2}z}{1 - K^{1/2}z}\right), \qquad K \geqslant 0;$$

$$h(z) = \frac{1}{|K|^{1/2}} \tan^{-1}(|K|^{1/2}z), \qquad K \leqslant 0. \tag{5.1}$$

Either of these expressions, when substituted into the general form (2.19) of the addition law, yields the result

$$w = (u + v)/(1 + Kuv). \tag{5.2}$$

The constant K cannot be negative, for if it were, the result of compounding two positive velocities both greater than $(-K)^{-1/2}$ would be a negative velocity. Positive velocities large enough to bring about this unsatisfactory state of affairs could always be attained by successive compoundings, since with K negative and w, u, and v all positive, w must *exceed* the sum of u and v.

If K is non-negative, however, it follows from (5.2) that $c(K) = K^{-1/2}$ is an invariant velocity: anything moving at this speed will have the same speed in all frames. The Galilean case is, of course, included in this range of possibilities, since $K = 0$ gives $w = u + v$, and $c(0) = \infty$. Historically, evidence for the nonzero value of K was provided by the growing body of evidence that the speed of light in empty space was in fact the invariant velocity.

It is not, however, necessary for there to be phenomena propagating at the invariant speed to reveal the value of K. One only requires sufficiently accurate measurements of the speed of any uniformly moving object from two different frames of reference. If the object C moves to the right in frame B, which itself moves to the right in frame A, then the addition law

$$v_{CA} = (v_{CB} + v_{BA})/(1 + Kv_{CB}v_{BA}) \tag{5.3}$$

can be used to express K in terms of the various velocities as

$$K = (1/v_{BA}v_{CA})[1 + x(1 - v_{CA}/v_{BA})], \tag{5.4}$$

where

$$x = v_{BA}/v_{CB} = -v_{AB}/v_{CB}. \tag{5.5}$$

Except for the quantity x, Eq. (5.4) expresses K entirely in terms of two velocities measured in the A frame. To determine the value of K, A-frame observers need therefore only measure the velocities v_{CA} and v_{BA} of the object C and the frame B, and collect from their associates in the B frame the value they have measured for the dimensionless ratio x. Since x does not depend on the time and distance units used by B-frame observers, it is not even necessary to cross-calibrate the B- and A-frame units. Equation (5.4) directly gives the inverse square of the invariant velocity in whatever units are employed in the A-frame.

Nor is light needed to establish in either frame a reliable method for measuring velocities. A and B could, for example, measure the velocities of each other and of C by setting up appropriate realizations of the *Gedanken* race of Sec. II, using the fact that in the rest frame of the race track Eq. (3.5) relates the measured fraction r to the speed s of the hare and speed u of the tortoise simply by

$$r = (s - u)/(s + u). \tag{5.6}$$

Thus A, by arranging a race between his own hare (the "A hare") and the object C, could measure r and determine the velocity u of C in terms of the speed s_A of the A hare. A similar determination could be made of the velocity of B and B's race track in units of s_A. B, on the other hand, would make measurements in terms of the B-hare velocity s_B. Because x is dimensionless, it would not be necessary to take care that B hares and A hares were prepared identically in their respective frames.

One can also dispense with the observers A and B. All that is really required are two superimposed parallel rigid race tracks, in uniform relative motion, each with its own hare, designed with a mechanism of propulsion that couples each hare to its track in the same way for the right-left and left-right parts of its race. By suitably lining everything up at the start and recording the fractional positions along each track where its hare first reencounters C and first reencounters the appropriate section of the other track, one can extract everything needed to give the value of K. No clock synchronization procedures – indeed, no clocks at all – are required.

Having demonstrated that there must be an invariant velocity, one can build up the rest of special relativity along more conventional lines. Alternatively, with the addition law at hand from the start, one can vary the development in any number of ways, avoiding, if one wishes, the use of anything that travels at the invariant speed.

A more interesting, instructive, and challenging exercise, however, is to try to refine those stages of the argument that I have already presented above. What, for example, is the simplest piece of apparatus necessary to extract K from the kinds of elementary counting procedures used in the *gedanken* race or the comparison of commensurate *gedanken* oscillators? What are the least assumptions about the nature of time, distance, and velocity necessary to give meaning to such procedures? How weak or natural can one make the smoothness assumptions?

There is, at a minimum, much good pedagogy to be had in the process of understanding the relativistic interconnections of space, time, and velocity, without the use of light. Working in the dark can be illuminating.

Acknowledgments

The analysis presented above evolved in the course of a correspondence with E. M. Purcell, who questioned why my argument in Ref. 6 had to make any use of light at all, and replied with patience and tact to my initially skeptical response. I am also indebted to him, and, through him, to C. Papaliolios for many observations and remarks that I have absorbed into my essay without, I hope, too much unbeneficial distortion. The paper itself was written in various hotel rooms and offices throughout Scandinavia. I am indebted to Nordita for paying the bills, and to Martti Salomaa, Gøran Grimvall, Anders Barany, Bengt Lundqvist, and C. J. Pethick for providing green Selectric typewriters on very short notice. Finally, I am indebted to Daniel Greenberger, for telling me about Terletskii's book.

Notes and references

1 A. Einstein, *Ann. Phys.* **17**, 891 (1905).

2 Another way to derive relativity without light is given by Y. P. Terletskii, *Paradoxes in the Theory of Relativity* (Plenum, New York, 1968), Sec. 7.

Terletskii shows that a general linear transformation between two sets of space–time coordinates must be of the Lorentz form, with c^2 an undetermined constant, if the inverse and products of all transformations are to have the appropriate structure. (This argument stands on its own, without the appeals to the dependence of mass on velocity with which Terletskii precedes and concludes it.) My approach is conceptually simpler (though analytically somewhat more intricate) because I require no comparison of space and time measurements in different frames to establish the general form of the velocity addition law; my argument is based on *gedanken* experiments that make no use of clocks at all.

3 Einstein begins Ref. 1 with the formulation of two postulates. The first, the principle of relativity, is that "dem Begriffe der absoluten Ruhe nicht nur in der Mechanik, sondern auch in der Elektrodynamik keine Eigenschaften der Erscheinungen entsprechen." (Not only in mechanics but also in electrodynamics the phenomena have no properties corresponding to the concept of absolute rest.) The second is that "sich das Licht im leeren Raume stets mit einer bestimmten, vom Bewegungszustande des emittierenden Körpers unabhängigen Geschwindigkeit V fortpflanze." (Light always propagates in empty space with a definite velocity c, independent of the state of motion of the emitting body.)

4 In this paper I only establish the existence of a velocity that is invariant in frames of reference moving parallel to that velocity. The extension to the other two spatial dimensions can be achieved by a somewhat lengthy but straightforward modification of the usual arguments used to develop the relativistic behavior of length and time measurements. To avoid distracting the reader from the central part of the argument and to keep the paper to a reasonable length, I have omitted this exercise.

5 One might object that such measurements make implicit use of light to establish a synchronization procedure for the clocks used to time the speeds. The methods I describe for determining K in Secs. III or IV, however, require no clocks. (Note, nevertheless, that there are methods for synchronizing distant clocks that do not rely on light. They can, for example, be synchronized by direct comparison at the point midway between their intended positions and then symmetrically transported to those positions.)

6 N. David Mermin, preceding essay. Rereading both arguments a few years later, it is clear to me that Sec. IV is so very much the nicer of the two, that I would urge the reader in a hurry to skip Sec. III entirely.

7 Equation (2.18) also follows directly from (2.17) and the condition $f(x, 0) = x$.

8 In the preceding essay the hare is called a photon and its speed c is asserted to be the same in all frames of reference. Here there is nothing special about the speed s of the hare.

9 At this point the argument of the preceding essay is virtually complete. One need only note that when $v = 0$, Eq. (3.7) reduces the expression (3.5) for r to $r = (s - u)/(s + u)$. But if $s = s_1 = s_2 = c$, then setting the two expressions for r equal to one another gives an equation for the only remaining unknown w, whose solution is the relativistic addition law. Here, however, s_1 and s_2 are

related to s through the very addition law we seek to determine, and we arrive not at the addition law, but at a functional equation it must satisfy. Further analysis is required to solve that equation.

10 The two balls can be regarded as the tortoise and hare of Sec. III, constrained now to have commensurate velocities, and required to keep running until they meet again at the original starting line.

Postscript

I received a number of letters after this paper appeared, including a surprising request to reprint it in the Soviet Union. Most of the letters politely (and sometimes impolitely) called to my attention a vast tradition of papers deriving special relativity without the second postulate, dating back at least to 1910. Somewhat abashed I reported this in the letter to the editor that appears below. My scholarship in the matter had clearly been deplorable. On the other hand none of my predecessors had used anything remotely like my argument, so as a physicist I remain unashamed.

Letter to the editor

I would welcome an opportunity to make amends for an entirely inadequate set of references to closely related work in my recent paper, "Relativity without light." Like many before me, I seem to have come upon the possibility of deriving the special theory without the second postulate entirely on my own (or rather accompanied only by those whose help I mention in the acknowledgments at the end of my paper). I did learn of the approach by Terletskii in time to add a reference, but I was unaware of a vast literature of which I mention only what I currently consider to be my most spectacular omissions:

(1) The point seems first to have been made in 1910 by Ignatowsky,[1] and shortly thereafter in 1911 by Frank and Rothe.[2]

(2) It is explicitly mentioned by Pauli, on p. 11 of his celebrated treatise[3] (though he does spoil the fun by adding that "nothing can, naturally be said about the sign, magnitude and physical meaning of α [my constant K]. From the group-theoretical

assumption it is only possible to derive the general form of the transformation formulae, but not their physical content.")

(3) The point has received considerable recent attention in the pages of this very journal, notably (but not exclusively) in the elegant treatments by Lee and Kalotas[4] and by Levy-Leblond.[5]

Many other papers could be cited.

My approach does, however, have a somewhat different character from those that have been called to my attention (to date!). To begin with it is more limited – I only consider motion along a single spatial dimension. Within this framework, however, I extract the relativistic velocity addition law without the second postulate directly from a simple *gedanken* experiment, without ever raising the question of how spatial and temporal coordinates or measurements in different frames of reference might be related. I therefore would have published precisely the same argument had I known of the earlier work, but I would have given the paper a different title, a different introduction, and a commendably lengthy list of references.

1 W. V. Ignatowsky, *Arch. Math. Phys. Lpz.* **17**, 1 (1910) and **18**, 17 (1911); *Phys. Z.* **11**, 972 (1910) and **12**, 779 (1911).
2 P. Frank and H. Rothe, *Ann. Phys. Lpz.* **34**, 825 (1911); *Phys. Z.* **13**, 750 (1912).
3 W. Pauli, in *Encyclopadie der mathematischen Wissenschaften*, V19 (B. G. Teubner, Leipzig, 1921), translated as *Theory of Relativity* (Pergamon, New York, 1958).
4 A. R. Lee and T. M. Kalotas, *Am. J. Phys.* **43**, 434 (1975).
5 J.-M. Levy-Leblond, *Am. J. Phys.* **44**, 271 (1976).

21

$E = mc^2$

(written with M. J. Feigenbaum)

The first argument relating the mass of a body to its energy content was given by Einstein in 1905.[1] In Einstein's *gedanken* experiment the change in energy content is produced by the emission of light and his analysis relies upon the transformation law for the energy of electromagnetic radiation. Since 1905 a variety of different arguments have been constructed deriving the relation without appealing to electrodynamics.[2] We would like to describe here a derivation that, while purely mechanical, extracts $E = mc^2$ by directly considering a single inelastic process very much like the one Einstein used in 1905. Our argument also brings out the important but sometimes obscured distinction between mass, the proportionality constant in the kinetic energy, on which $E = mc^2$ has a direct and profound bearing, and mass, the rest energy divided by c^2, for which $E = mc^2$ is little more than a convenient convention.

We first review Einstein's 1905 argument. Consider a particle at rest with energy content (*Energieinhalt*) E_1, that emits in opposite directions two identical quantities of light (*Lichtmenge*), each with energy ε. After the emission the particle remains at rest with energy content E_2. Energy conservation requires that

$$E_1 = E_2 + 2\varepsilon. \tag{1}$$

Consider next the same decay in a frame of reference moving with velocity v along a direction making an angle θ with the direction of propagation of the emitted light. Call the initial and final energies of the particle in that frame $E_1(v)$ and $E_2(v)$. Einstein knew from electrodynamics how the energies of the two quantities of light transformed, and therefore he knew that energy conservation in the moving frame required

$$E_1(v)=E_2(v)+\frac{\varepsilon(1-v\cos\theta/c)}{\sqrt{1-v^2/c^2}}+\frac{\varepsilon(1+v\cos\theta/c)}{\sqrt{1-v^2/c^2}}. \qquad (2)$$

Subtraction of (1) from (2) gives

$$[E_1(v)-E_1]-[E_2(v)-E_2]=2\varepsilon(1/\sqrt{1-v^2/c^2}-1). \qquad (3)$$

But $E(v)-E$ (with either subscript) is just the difference in the energy of a particle moving with speed v and the energy of the same particle when at rest, i.e., it is the kinetic energy of the particle $K(v)$ which, when v is small compared to c, is just $\frac{1}{2}mv^2$. In this same limit the right-hand side of Eq. (3) simplifies to $\varepsilon v^2/c^2$, and therefore Eq. (3) informs us that

$$m_1-m_2=2\varepsilon/c^2. \qquad (4)$$

Therefore, notes Einstein, if a body gives off a quantity of energy in the form of radiation, then its mass diminishes by that energy divided by c^2. He then remarks with characteristic boldness that it makes no difference that the energy withdrawn from the body happens to have gone into radiation. This leads him to conclude that if the energy of a body changes for any reason, then its mass changes by the amount of the energy change divided by c^2.

Although his conclusion is independent of electrodynamics, Einstein had to invoke it at an intermediate step to determine how the energy of the radiation changed with a change of frame. We, being more at home with relativistic kinematics than Einstein's audience in 1905, can use virtually the same *gedanken* experiment to reach his conclusion without ever leaving the realm of mechanics. The knowledge of how the energy of radiation transforms is replaced by the velocity addition law, which is most conveniently expressed in the following form:

Let a body move uniformly with speed u. Then in a frame of reference moving with speed v at an angle θ to the direction of motion of the body, its speed u' satisfies[3]:

$$\frac{1}{\sqrt{1-u'^2/c^2}}=\frac{1-uv\cos\theta/c^2}{\sqrt{1-u^2/c^2}\sqrt{1-v^2/c^2}}. \tag{5}$$

Equipped with this addition law, we replace the two identical quantities of light in Einstein's argument by two identical particles, about which we know no more than we do about the particle originally under investigation. We thus consider a particle at rest with energy content E_1, that emits two identical particles, moving in opposite directions with the same speed u. After the emission the emitting particle remains at rest (as symmetry requires) with energy content E_2. The energy of each of the emitted particles is some function of its speed $E_3(u)$, the form of which is to be determined. Energy conservation requires that

$$E_1-E_2=2E_3(u). \tag{6}$$

Following Einstein, we next consider the same decay in a frame of reference that moves with speed v at an angle θ to the direction of emission of the particles. The initial and final energies of the emitting particle in that frame are $E_1(v)$ and $E_2(v)$. The emitted particles will have energies $E_3(u')$ and $E_3(u'')$, where u' and u'' are given by Eq. (5) with the two different choices of sign for $\cos\theta$. Energy conservation in the new frame requires that

$$E_1(v)-E_2(v)=E_3(u')+E_3(u''). \tag{7}$$

This relation must hold whatever the angle θ between the direction of motion of particles 3 and the new frame of reference. Since the left-hand side of Eq. (7) is independent of θ, we conclude that the quantity

$$E_3(u')+E_3(u'') \tag{8}$$

cannot depend on θ.

Note that in Einstein's derivation this independence of θ is an immediate consequence of the transformation rules for the quantities of light and is explicit in Eq. (2). Indeed it is quite unnecessary to consider general θ to establish Einstein's argument (although he did); the case $\theta = 0$ by itself leads directly to the result. In our case, however, the requirement of θ independence imposes a strong constraint on the unknown function $E_3(u)$, which turns out to be enough to restore the ground lost by renouncing the use of electrodynamics. The argument is as follows:

Since the speed of particle 3 cannot exceed[4] c, we can equally well consider $E_3(u)$ to be a function of the variable $1/\sqrt{1 - u^2/c^2}$:

$$E_3(u) = f(1/\sqrt{1 - u^2/c^2}). \tag{9}$$

This change of variable is designed to take advantage of the fact that $1/\sqrt{1 - u'^2/c^2}$ and $1/\sqrt{1 - u''^2/c^2}$ depend linearly on $\cos\theta$ in the velocity addition law (5). Thus $E_3(u') + E_3(u'')$ becomes

$$f\left(\frac{1 - uv\cos\theta/c^2}{\sqrt{1 - u^2/c^2}\sqrt{1 - v^2/c^2}}\right) + f\left(\frac{1 + uv\cos\theta/c^2}{\sqrt{1 - u^2/c^2}\sqrt{1 - v^2/c^2}}\right). \tag{10}$$

This will evidently be independent of $\cos\theta$ if f is a linear function of its argument. More importantly, a little elementary analysis shows that if f is a continuous function of its argument (as an acceptable energy function must surely be) then Eq. (10) can be independent of $\cos\theta$ for frames moving with arbitrary speeds v *only* if f has the linear form $f(z) = az + b$.[5] This is our central mathematical result.

But if f has this form, then the energy of the particles 3 must depend on their velocity according to the law

$$E_3(u) = E_3 + k_3(1/\sqrt{1 - u^2/c^2} - 1), \tag{11}$$

where E_3 and k_3 are velocity-independent constants characteristic of particles of type 3.

The constant $E_3 = E_3(0)$ is the energy content of either of the particles of type 3 in its rest frame, and the constant k_3 determines the overall scale of its kinetic energy,

$$K_3(u) = E_3(u) - E_3 = k_3(1/\sqrt{1 - u^2/c^2} - 1). \tag{12}$$

Since no special properties of particles 3 entered into the above argument, we may conclude that any type of particle must be characterized by constants E and k such that the velocity dependence of its energy is given by

$$E(u) = E + k(1/\sqrt{1 - u^2/c^2} - 1). \tag{13}$$

In particular, using the form (13) for the emitting particle (with subscripts 1 and 2 for its pre- and post-emission states) we can write the equation of energy conservation in the moving frame (Eq. (7)) as

$$E_1 + k_1\left(\frac{1}{\sqrt{1 - v^2/c^2}} - 1\right) - E_2 - k_2\left(\frac{1}{\sqrt{1 - v^2/c^2}} - 1\right)$$
$$= 2E_3 + 2k_3\left(\frac{1}{\sqrt{1 - u^2/c^2}\sqrt{1 - v^2/c^2}} - 1\right). \tag{14}$$

By subtracting from Eq. (14) its form in the original ($v = 0$) frame, and canceling a factor of $1/\sqrt{1 - v^2/c^2} - 1$ common to all terms, one extracts the following relation between the kinetic energy coefficients of the emitting particle before and after the emission:

$$k_2 = k_1 - 2k_3/\sqrt{1 - u^2/c^2} = k_1 - 2k_3 - 2K_3(u). \tag{15}$$

This is our central physical result.

By examining the u dependence of the kinetic energy (12) when u is small compared with c we learn that in the nonrelativistic limit k/c^2 is called the mass m of the particle. If we adopt that traditional nomenclature, the content of (15) is that m_2 is less than $m_1 - 2m_3$ by $1/c^2$ times the kinetic energy with which the type 3 particles are expelled in the rest frame of the emitting particle. This is Einstein's conclusion, modified by the fact that in our version of the argument the emitted particles carry away mass as well as kinetic energy.

Thus if a particle in its rest frame decays into a collection of particles with a total kinetic energy K, then the total mass of the final group of particles must be less than the initial mass by K/c^2. Note that the mass that enters into this relation is the sum of kinetic energy coefficients k/c^2. If a particle gives up mass to produce kinetic energy K in such a process, then this loss is directly reflected in a lowering of the energy needed to accelerate the particle to a given velocity, as determined by its kinetic energy coefficient k. It is a remarkable fact that the simple relativistic kinematics of space and time measurements embodied in the addition law (5), imply by themselves that if energy is to be conserved at all, then it is necessary that the relation between the velocity and kinetic energy of a particle be subject to this kind of modification.

We emphasize the profound character of this conclusion, because the rest energy E in Eq. (13), which has not yet entered into the argument at all, is also conventionally assigned the numerical value mc^2. In contrast to the remarkable relation (15) obeyed by the kinetic energy coefficient, this other use of $E = mc^2$, though it is sometimes cited with comparable fanfare, has very little content. To see this we examine further the implications of energy conservation for the rest energies E.

Consider a group of particles, initially moving with velocities u_i with total energy

$$U = \sum_i E_i(u_i) = \sum_i \left(g_i + \frac{k_i}{\sqrt{1 - u_i^2/c^2}} \right), \tag{16}$$

$$g_i = E_i - k_i. \tag{17}$$

After some number of collisions, emissions, or mergers, the velocities, kinetic energy coefficients k_i, and the rest energies E_i may all have changed, but only in such a way that the total energy U remains unaltered. This conservation law must hold in all frames of reference. Since the k_i and E_i (and hence the g_i) are all invariant under a change of frame, only the velocities u_i depend on frame of reference.

The total energy U' in a frame moving with velocity v is therefore simply given by applying the addition law (5) to the velocities in Eq. (16):

$$U' = \sum_i \left(g_i + k_i \frac{1 - \mathbf{u}_i \cdot \mathbf{v}/c^2}{\sqrt{1 - u_i^2/c^2}\sqrt{1 - v^2/c^2}} \right). \tag{18}$$

If U and U' are both conserved, then so is $U - U'\sqrt{1 - v^2/c^2}$. But this last quantity is just

$$(1 - \sqrt{1 - v^2/c^2}) \sum_i g_i + \mathbf{v} \cdot \sum_i \frac{(k_i/c^2)\mathbf{u}_i}{\sqrt{1 - u_i^2/c^2}}. \tag{19}$$

The quantity (19) must be conserved whatever the velocity $\mathbf{v}$ of the moving frame. Since the first sum is even in $\mathbf{v}$ and the second odd, both must separately be conserved. The second sum is nothing but the relativistic momentum (with, note well, the kinetic energy coefficient of each particle divided by c^2 playing the role of its mass m); the first is just the sum of all the g_i.

We have therefore deduced (in the conventional way) that energy conservation implies momentum conservation, but also the conservation of something else. What are we to make of this? Note first that since $g = E - k$ is additively conserved, then because the additively conserved quantity $E(u)$ has the structure (13), the quantity $k/\sqrt{1 - u^2/c^2}$ is itself additively conserved, and as good a candidate for the energy function as the more general form (13). We can take advantage of this to redefine the energy function to be given by

$$E(u) = k/\sqrt{1 - u^2/c^2}, \tag{20}$$

but we must then keep in mind that if there is an additively conserved Lorentz invariant associated with each particle, then E can be trivially redefined to include an additional velocity-independent term proportional to that quantity, without doing violence to the conservation laws.

This is as far as one can go without a more detailed dynamical theory. Electric charge, for example, is such a Lorentz invariant additively conserved quantity. So (except in grand unified theories) is baryon or lepton number. At the level on which we are operating there are no logical grounds *a priori* for excluding terms proportional to such quantities in the definition of rest energy. There is, however, something to be said for taking g to be zero in the definition of $E(u)$, resulting in the definition (20). For the choice $g = 0$ identifies the rest energy E of a particle with its kinetic energy coefficient k. The rest energy then becomes a measure of how much energetic capital a particle has available for the production of new kinetic energy. The conceptual price one pays for this convenience is the blurring of the distinction between the rest energy and the kinetic energy coefficient.

Note, finally, what happens when one tries to make the same argument directly from the nonrelativistic addition law $u'^2 = u^2 + v^2 - 2uv\cos\theta$. One immediately deduces the nonrelativistic version of Eq. (13): $E(u) = E + \frac{1}{2}mu^2$ (giving the kinetic energy coefficient its conventional name). The crucial result (15), however, is lost. One finds only $m_2 = m_1 - 2m_3$. Using this one deduces that $E_2 = E_1 - 2(E_3 + \frac{1}{2}m_3u^2)$, i.e., that the extra kinetic energy in the inelastic process must be subtracted from the rest energy of particle 2. In the nonrelativistic case this is nothing but a trivial bookkeeping device – the limit of the trivial relativistic convention about the rest energy. What is lacking in the nonrelativistic argument is any basis for making the nontrivial link between the creation of kinetic energy and the dynamically important kinetic energy coefficients, that gives $E = mc^2$ its deep physical content.

Appendix A

Let the 4-velocity of the emitted particle be

$$u^{(4)} = (c,\mathbf{u})/\sqrt{1 - u^2/c^2}, \tag{A1}$$

and the 4-velocity of the moving frame,

$$v^{(4)} = (c,\mathbf{v})/\sqrt{1 - v^2/c^2}, \tag{A2}$$

The Lorentz invariant inner product of these is

$$u^{(4)} \cdot v^{(4)} = \frac{c^2 - \mathbf{u}\cdot\mathbf{v}}{\sqrt{1 - u^2/c^2}\sqrt{1 - v^2/c^2}}. \tag{A3}$$

The value of this quantity is independent of frame. But in the moving frame v is 0 and u has, by definition, the value u', so that

$$u^{(4)} \cdot v^{(4)} = c^2/\sqrt{1 - u'^2/c^2}. \tag{A4}$$

The addition law (5) is simply the statement that the evaluations (A3) and (A4) of the inner product are the same.

Appendix B

One can prove quite simply that the energy function f is linear by assuming that the dependence of energy on velocity is smooth enough to be twice differentiable. (In Appendix C we relax this to a simple assumption of continuity.) Since the energy (10) is independent of the angle θ, its second θ derivative (at fixed u and v) must vanish. This gives

$$f''\left(\frac{1-uv\cos\theta/c^2}{\sqrt{1-u^2/c^2}\sqrt{1-v^2/c^2}}\right) + f''\left(\frac{1+uv\cos\theta/c^2}{\sqrt{1-u^2/c^2}\sqrt{1-v^2/c^2}}\right)=0. \quad \text{(B1)}$$

Setting $\cos\theta=0$ in (B1) and noting that the speed v of the moving frame can have any value between 0 and c, we learn that

$$f''(z)=0, \qquad z\geqslant 1/\sqrt{1-u^2/c^2}. \quad \text{(B2)}$$

Since u is fixed[6] Eq. (B2) by itself is not enough. To establish that f'' vanishes over the rest of its range set $\cos\theta=1$ in Eq. (B1). With u and v positive, the second term in Eq. (B1) then vanishes as a consequence of Eq. (B2), and therefore so must the first:

$$f''\left(\frac{1-uv/c^2}{\sqrt{1-u^2/c^2}\sqrt{1-v^2/c^2}}\right)=0. \quad \text{(B3)}$$

As v varies from 0 to u, the argument of f'' in (B3) goes from $1/\sqrt{1-u^2/c^2}$ down to unity, giving us the vanishing of f'' in the rest of its range. It follows that throughout its entire range $1\leqslant z<\infty$, f must be of the form $az+b$.

Appendix C

With a little more effort (and without the use of calculus), one can reach the conclusion of Appendix B, making only the weaker (and physically even more compelling) assumption that the energy function f is continuous. If the energy (10) is independent of $\cos\theta$ then it is equal to its value when $\cos\theta=0$:

$$f\left(\frac{1-uv\cos\theta/c^2}{\sqrt{1-u^2/c^2}\sqrt{1-v^2/c^2}}\right)+f\left(\frac{1+uv\cos\theta/c^2}{\sqrt{1-u^2/c^2}\sqrt{1-v^2/c^2}}\right) = 2f\left(\frac{1}{\sqrt{1-u^2/c^2}\sqrt{1-v^2/c^2}}\right). \quad \text{(C1)}$$

We must show for fixed u, that if this condition holds for $0 \leqslant v/c < 1$ and $-1 \leqslant \cos\theta \leqslant 1$, then $f(z)$ must be linear in z for all $z \geqslant 1$.

Note first that it is enough to establish that $f(z)$ is linear when $z \geqslant 1/\sqrt{1-u^2/c^2}$. For with $\cos\theta$ equal to unity, the arguments of the second and third occurrences of f in (C1) are both greater than or equal to $1/\sqrt{1-u^2/c^2}$; replacing f in both cases by $az+b$, one finds that

$$f(x) = ax + b, \qquad x = \frac{1 - uv/c^2}{\sqrt{1-u^2/c^2}\sqrt{1-v^2/c^2}}. \tag{C2}$$

But as v goes from 0 to u, x goes from $1/\sqrt{1-u^2/c^2}$ to 1, thereby establishing that f has the form $az+b$ in the rest of its range.

Next define

$$f\left(\frac{t}{\sqrt{1-u^2/c^2}}\right) = g(t). \tag{C3}$$

We have

$$g\left(\frac{1 - uv\cos\theta/c^2}{\sqrt{1-v^2/c^2}}\right) + g\left(\frac{1 + uv\cos\theta/c^2}{\sqrt{1-v^2/c^2}}\right) = 2g\left(\frac{1}{\sqrt{1-v^2/c^2}}\right), \tag{C4}$$

and must show that $g(z)$ is linear for $z \geqslant 1$. With new variables

$$x = 1/\sqrt{1-v^2/c^2}, \qquad y = (uv/c^2)\cos\theta/\sqrt{1-v^2/c^2}, \tag{C5}$$

Eq. (C4) becomes

$$g(x) = \tfrac{1}{2}[g(x+y) + g(x-y)] \tag{C6}$$

for all x and y satisfying

$$x \geqslant 1, \qquad -(u/c)\sqrt{x^2-1} \leqslant y \leqslant (u/c)\sqrt{x^2-1}. \tag{C7}$$

The inequalities (C7) confine x and y to the interior of the right-hand branch of a hyperbola. Consider any square within this region, centered at a point on the x axis, and tilted at 45° so that diagonally opposite vertices of the square meet the axis at $x = a$ and $x = b$ (Fig. 1). If we can show for any such square that $g(x)$ is linear for $a \leqslant x \leqslant b$, then we shall have established that g is linear for all $x \geqslant 1$, since the entire x axis beyond $x = 1$ can be covered by overlapping squares of this kind. Since the sides of the square are

$x \pm y = a$, $x \pm y = b$, if (C6) holds for x, y in the square, then with $x + y = t$, $x - y = s$, we must also have

$$g[\tfrac{1}{2}(r + s)] = \tfrac{1}{2}[g(r) + g(s)], \qquad \text{(C8)}$$

for all r and s in the interval $[a, b]$.

Equation (C8) asserts that the value of g midway between *any* two points in the interval $[a, b]$ is just the linear interpolation of the value at those two points. Consequently, if we have a set of points S from $[a, b]$ on which g is linear, then g will remain linear on S when we add to it the points midway between any two of the original points. Exploiting this fact, we can easily construct a dense set S of points in $[a, b]$ on which g is linear. We first put into S the two endpoints of the interval. We can then add to S the midpoint of the interval. Having done this we can add the remaining points that divide the interval into quarters, since they are halfway between the two ends and the midpoint. In the same way, we can next add to S the remaining points that divide the interval into eighths, then sixteenths, thirty-seconds, etc., all the while maintaining the linearity of g on S. In this way we will eventually add to S points arbitrarily close to any point in $[a, b]$.

Fig. 1. The region in Eq. (C7) lies within the right-hand branch of the hyperbola $x^2 - (c/u)^2 y^2 = 1$. The square shown within that region contains all points (x, y) with $a \leqslant x \pm y \leqslant b$.

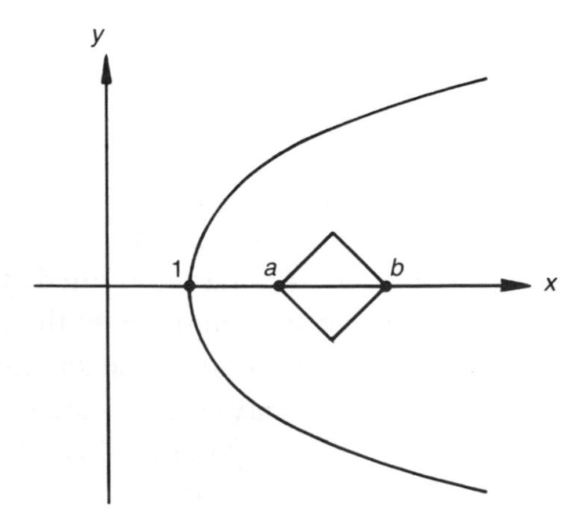

Therefore if g is continuous in $[a, b]$ it must be linear on the entire interval.

Notes and references

1 A. Einstein, *Ann. Phys.* **18**, 639 (1905).

2 For a critical discussion of many examples, see the references in P. C. Peters, *Am. J. Phys.* **54**, 804 (1986). Peters himself gives a discussion of the relativistic conservation laws that, like ours, derives and solves a functional equation for the form of the unknown conserved quantities. Peters finds the standard relativistic expressions by considering an elastic collision. Assuming that these forms continue to be conserved in inelastic collisions he then derives, along traditional lines, the dependence of mass on energy content.

3 See Appendix A for a particularly simple derivation of Eq. (5).

4 We appeal to causality; an appeal to dynamics would introduce a disquieting circularity into the argument. Note that with this change of variables Einstein's version of the *gedanken* experiment (with $u = c$) can only be viewed as a singular limiting case of ours.

5 This is done in Appendix B under the stronger assumption that the energy function f is smooth enough to be twice differentiable. In Appendix C we give a slightly more elaborate argument (suggested by C. L. Henley) that requires only continuity (and has the further virtue of being accessible, if presented with care, to students innocent of the calculus.)

6 In an earlier version of this argument, we allowed u to vary, but assumed that the particles 3 could be emitted in the same internal state whatever the value of the speed u at which they emerged. We are indebted to E. M. Purcell for pointing out the restrictive nature of this assumption, and questioning the need for it. The present argument requires only a single speed of emission, restoring the necessary flexibility by introducing a continuum of directions for the moving frame.

IV.

Mathematical musings

22

Logarithms!

Introduction

Among other things, this is indeed a paper about Logarithms! I hope the difference between Logarithms! and logarithms will be clear at the end.[1] Evidently Logarithms! are the nicer kind, but if that is not enough to entice you to read on, let me at once answer three questions:

(1) For whom am I writing?

(2) What about?

(3) Why?

(1) I am writing for physicists who teach the one-year course in "liberal arts physics" for students with absolutely no trace of professional interest in the subject. Although the paper itself simply presents the contents of the lectures on Logarithms! that I have given such students for several years, the presentation here is aimed at a more sophisticated audience of colleagues, to reveal to them briefly and efficiently how the techniques can be used. The paper might be of some interest to the students themselves, but if they can follow the argument without help they probably don't belong in the course for which it is intended.

(2) I am writing about three different things:

(a) An efficient and entertaining way to review arithmetic skills in manipulating powers, which, for typical students in the "liberal arts" physics course, may have lain unused for as many as four years. Without restoring such skills to them it is virtually impossible to teach any physics whatever.

(b) The attitude of physicists toward numbers. I believe this is one of the most important points that can be put across in a physics course for liberal arts students. A number unaccounted for cries out for explanation, and the act of explaining is an act of conquest. The number explained is a trophy to be treasured. Needless to say nothing this blatant is said explicitly, either in my own course or in the text that follows, but the attitude is made plain enough to most students in the course of the discussion.

(c) An introduction, by example, to the art of estimating and a demonstration of its remarkable power. Many students have never been taught that you can estimate anything, without losing dignity. In my physics course we estimate things all day long, and the confidence gained by starting the estimating on known numbers is important. By furnishing us all with portable computers, the electronics industry has deprived us of many sources of insight and pleasure.

(3) I am writing about these matters because they bear directly on the interests and needs of college and university physics teachers and students, at the level of the "liberal arts physics" course. If calculus is the mathematical language of the introductory course for physics majors, and algebra is the language of the introductory course for premedical students, then arithmetic is the language of liberal arts physics. Students quite impervious to the usual algebraic treatment of the relativity of simultaneity, can be carried remarkably far if only the case $v/c = 3/5$ is studied, for much of the distracting analytic baggage underlying the derivation of the Lorentz transformation vanishes away when letters are replaced by numbers. In a similar way, many students become frightened or bored, or simply refuse to listen, when "reminded" that $a^{b/c}$ is the cth root of a product of b a's. Wrestling with log 2 for a week convinces many of them that it is, after all, simple, useful, and even fun.

It would be an act of downright silliness on my part to try to enumerate the areas where a knowledge of even liberal arts physics is hampered by an almost phobic refusal to focus on $a^{b/c}$. At least two distinguished physics texts devote several pages to the computation of base 10 logarithms, for much the same pedagogic reasons as I have outlined.[2] The approach I shall describe succeeds in being much less arithmetically cumbersome than these, by replacing a systematic but somewhat turgid approach by an *ad hoc* blend of artistry and good fortune. But it belongs to the same tradition of physics pedagogy and should find similar applications.

Logarithms!

The numerical explorations to be described below were developed in the course of reviewing the elementary properties of positive, negative, and fractional exponents that inevitably arise, even in physics courses for nonscientists. Students with astonishingly slight capacities for algebraic abstraction have found these constructions entertaining and instructive, those with a little taste for numbers have found them intriguing, and, indeed, I myself have found them a recurring obsession, which this article is written in the hope of curing.

If you consult a ten place table of base 10 logarithms, or the kind of pocket calculator the nonscientist is unlikely to have access to, you will find that

$$\log 2 = 0.3010299956\ldots. \tag{1}$$

Everybody knows (or at least knew, before they all had pocket calculators) that with much information of this kind one can reduce the odious task of multiplying two long numbers to the merely distasteful one of adding two others, and a little hunting through tables. By having many sunny hours of our adolescence filled with such pursuits, some of us have acquired a deep hatred for logarithms, and even for the elementary properties of positive, negative, and fractional exponents from which they derive.

To overcome this antipathy, while illustrating and applying the elementary properties, I suggest that one should inquire not into the uses of Eq. (1), but into its very origins. Why should anyone believe that to eight place accuracy the logarithm of 2 is 0.30103?

We approach the question in stages. It does not take long to figure out that

$$2^{10} = 1024, \tag{2}$$

a number which, by sheer chance, is very little different from

$$10^3 = 1000. \tag{3}$$

Ignoring the difference, we then have

$$2^{10} \approx 10^3, \tag{4}$$

so that

$$2 \approx 10^{3/10} = 10^{0.3} \tag{5}$$

which tells us that

$$\log 2 \approx 0.3, \tag{6}$$

a pretty good guess. Since the left side of Eq. (4) is actually a little bigger than the right side, we know that log 2 is somewhat larger than 0.3, and Eq. (1) begins to make a little sense.

But how much bigger than 0.3? How much of that long tail can one come to terms with? There are two ways to proceed: by systematic analysis, or by hoping for more lucky accidents of the kind that gave us so good an estimate for log 2 (missing, after all, by only a third of a percent). The second method is the more entertaining, and we pursue it first.

Equations (2) and (3) give about as good a coincidence as you can hope for, if you want a simple approximate relation involving only powers of 2 and 10, and accept the ground rules that everything should be calculable with a pencil and one small sheet of paper. But if we let a few more integers into the game, relations at least as good as (4) (which makes about a 2.5% error) lie all about. The only integers whose logarithms need concern us are the prime numbers, since the logarithms of composites are the sums of those of their prime factors. We therefore inquire into the logarithms of 3, 7, and 11, as well as 2. (Since $\log 5 = 1 - \log 2$, its determination is not of separate concern.) The data we wish to account for are:

$$\log 3 = 0.4771212547\ldots, \tag{7}$$

$$\log 7 = 0.8450980400\ldots, \tag{8}$$

$$\log 11 = 1.041392685\ldots. \tag{9}$$

Let us first estimate these by finding some 1% or 2% coincidences like the one that gave us 3/10 for log 2. By making use of that first estimate, the search for coincidences is not a hard one. Thus

$$3^4 = 81, \qquad 2^3 \times 10 = 80, \tag{10}$$

so with little more than a one percent error,

$$3^4 \approx 2^3 \times 10, \tag{11}$$

or

$$\log 3 \approx (3 \log 2 + 1)/4 \approx 19/40 = 0.475, \tag{12}$$

which misses (7) by only $\frac{1}{2}$%. In a similar vein, since

$$7^2 = 49, \qquad 10^2/2 = 50, \tag{13}$$

with the estimate of 3/10 for log 2, we find

$$\log 7 \approx 17/20 = 0.85, \tag{14}$$

which is less than a percent away from Eq. (8). Using the estimates (12) and (6), and noting that

$$11^2 = 121, \qquad 2^2 \times 3 \times 10 = 120, \tag{15}$$

we find

$$\log 11 \approx 83/80 = 1.0375, \tag{16}$$

which is less than $\frac{1}{2}$% away from Eq. (9).

Those whom the schools have taught to hate numbers will probably want to stop after this preliminary warmup, but the real fun is only beginning. For if we look a little longer at numbers whose prime factors are only powers of 2, 3, 5, 7, and 11, we uncover some truly splendid coincidences. I begin with my prize specimen:

$$3^4 \times 11^2 = 9801, \qquad 2 \times 7^2 \times 10^2 = 9800. \tag{17}$$

I must confess to having made this discovery while leafing through a table of prime factors up to 10 000. Once found, of course, it is easily verified, but I felt the search that brought it to light to be offensively at odds with the back of the envelope spirit of the whole enterprise. My mind was set at rest when I noticed, without anybody's help, that 98, 99, and 100 are *three* consecutive numbers composed only of primes less than or equal to 11, and Eq. (17) is merely the expression of the fact that $100 \times 98 = 99^2 - 1$.

Equation (17) gives us an 0.01% coincidence, and therefore a much improved linear relation between the logarithms of 2, 3, 7, and 11. Since this is one equation in four unknowns, we need three more independent coincidences of comparable accuracy to solve for the logarithms. Counting up from 100 until boredom set in, I came upon no additional trios of consecutive numbers composed only of the primes 2, 3, 5, 7, and 11. Counting downwards, however, one encounters the groups [54, 55, 56], [48, 49, 50], and [20, 21, 22]. These yield

$$10^2 \times 11^2/2^2 = 3025, \qquad 2^4 \times 3^3 \times 7 = 3024 \tag{18}$$

$$7^4 = 2401, \qquad 2^3 \times 3 \times 10^2 = 2400 \tag{19}$$

$$3^2 \times 7^2 = 441, \qquad 2^2 \times 10 \times 11 = 440. \tag{20}$$

Equations (18) and (19), while not up to the standard set by Eq. (17), still give 0.03% and 0.04% coincidences. The estimates they furnish are more than an order of magnitude better than those of our first round. Equation (20), however, while still substantially better than the best of the first round coincidences, is unquestionably the weak link in the chain. I would be delighted to learn of something simple and better. But it is not helpful to note that $2 \times 3^7 = 4374$, while $7 \times 10^4/2^4 = 4375$. The resulting equation for the logarithms is not linearly independent of those coming from Eqs. (17) and (18). Equally useless, for similar reasons, is the fact that $2^2 \times 3^6 \times 7^3 = 1\,000\,188$. I mention this last one not merely to show off, but also to indicate that I did struggle a bit before settling for Eq. (20).

If one ignores the slight differences in the pairs of numbers appearing in Eqs. (17)–(20), the four simple linear equations for the four logarithms are easily solved and give:

$$\log 2 \approx \frac{72}{239} = 0.30125\ldots \quad \text{(exact: } 0.30102\ldots\text{)}, \tag{21}$$

$$\log 3 \approx \frac{114}{239} = 0.47698\ldots \quad \text{(exact: } 0.47712\ldots\text{)}, \tag{22}$$

$$\log 7 \approx \frac{202}{239} = 0.84518\ldots \quad \text{(exact: } 0.84509\ldots\text{)}, \tag{23}$$

$$\log 11 \approx \frac{249}{239} = 1.04184\ldots \quad \text{(exact: } 1.04139\ldots\text{)}, \tag{24}$$

Off by no more than 2 in the 4th significant place!

About half the class may feel some joy at this. The rest will be bored no matter what happens next, so you may as well go on to point out that one can, with only a little more effort do better. Much better!

The search for bigger and better coincidences is about over if one wishes to pursue it by amateur means. I offer with a slight blush

$$2^{53} = 9.007199\cdots \times 10^{15} \tag{25}$$

and

$$2 \times 3^{35} = 1.0006309\cdots \times 10^{17}, \tag{26}$$

which I admit to having discovered with the aid of a calculator. Not only does this method violate the rules for amateurs. It considerably undermines one's stand against students who might wonder sullenly why you do not ask a calculator to tell you the wretched logarithms in the first place. Most students crass enough to feel this way, however, have probably never been exposed to calculators that handle logarithms, but only to those that multiply, so the danger is not great. However a more honorable defense can be constructed around the fact that these relations are not as hard to verify by hand as one might think. Thus

$$2^{53} = 8 \times (1024)^5 = 8 \times 10^{15} \times (1 + 24/1000)^5, \tag{27}$$

from which an application of the binomial theorem (known to many amateurs) easily extracts Eq. (25) to the required accuracy. The other relation (26) requires a little more effort, but fits with little trouble onto a well organized page. The resulting approximations,

$$2^{53} \approx 3^2 \times 10^{15}, \qquad 2 \times 3^{35} \approx 10^{17}, \tag{28}$$

give two linear equations which yield:

$$\log 2 \approx \frac{559}{1857} = 0.3010231\ldots \quad \text{(exact: 0.3010299\ldots),} \tag{29}$$

$$\log 3 \approx \frac{886}{1857} = 0.4771136\ldots \quad \text{(exact: 0.4771212\ldots).} \tag{30}$$

The disagreement is down to less than one in the 5th significant place.

To improve on this we must invoke a small consequence of systematic analysis:

When X is a number (positive or negative) small in size compared with one, then the logarithm of $1 + X$ is very accurately given by X itself, multiplied by the number 0.43429...:

$$\log(1 + X) \approx 0.4343X, \quad \text{for } |X| \ll 1. \tag{31}$$

The smaller the size of X, the more accurate the approximation. The number 0.43429... (which professionals will recognize as the base 10 logarithm of e) can be truncated to 0.4343 with two gratuitous benefits: the accuracy is virtually to five places (rather than the apparent four), and tender minds need only remember the number 43 (and then again 43) rather than a full four digit number.

Persuading the amateur that there is some truth in Eq. (31) can be arduous. There is something to be said for simply producing it as one might a laser: a technological spin-off of higher research, with spectacular capabilities, now to be shown. For the more thoughtful students, however, one can first observe that since the logarithm of one is zero, when X is a very small number the logarithm of $1 + X$ can also be expected to be a very small number. The question is why that second very small number should be simply proportional to the first. Once can next point out that when Y and Z are both very small, then the product $(1 + Y)(1 + Z)$ is equal to $1 + (Y + Z)$ except for a correction that is *very* very small, and therefore the logarithm of $1 + (Y + Z)$ must be *very* very close to the sum of the logarithms of $(1 + Y)$ and $(1 + Z)$. Anyone can see that $\log(1 + X) \approx AX$, A a constant, has this necessary property, and with somewhat more squirming one can even be fairly persuasive on the point that nothing else does.

This leaves only the sticky question of why the constant A should be 0.43429.... There was a time in the evolution of this pedagogical adventure, when I found it necessary to break the ground rules by producing a list of successive square roots of 10, and noting that the values of

$$A_n = \frac{1}{2^n(10^{1/2^n} - 1)} \tag{32}$$

settled down to 0.4343 with gratifying efficiency. I lived with this departure from the back of the envelope by noting to myself that the calculator purchased by the amateur at the corner drugstore often gives square roots, but rarely logarithms. Then one day it occurred to me that such inferior instruments would be more than likely to yield round off errors which would do the amateur in before that second 43 had a chance to be established. With this, my conscience rebelled, and out of another appeal to the binomial theorem and no more than the usual amount of abstract mumbo jumbo, I constructed a computation in the purest amateur spirit, which readers who persevere to the end of this essay are invited to examine in the Appendix.

Armed with the approximation (31), we can improve our efforts at all stages. At the crudest level we replace the estimate (4) by the exact identity:

$$2^{10} = 10^3 \times (1 + 24/1000). \tag{33}$$

Since 24/1000 is a small number, we can use Eq. (31) to estimate with some accuracy the size of the logarithm of the correction factor, and with no more additional effort than that required to multiply 24 by 0.4343 we arrive at the improved value:

$$\log 2 \approx 0.301042\ldots \quad \text{(exact: } 0.301029\ldots\text{)}. \tag{34}$$

This is about as good as the result (29) of the most extreme extension of the coincidence method, and will convince those students who still have a taste for the game, of the marvelous power of the 0.4343 method.

If one similarly improves the rough estimates furnished by Eqs. (10), (13), and (15), using the values 1/80, $-1/50$, and 1/120 for X, all of which the amateur can multiply by 0.4343 through an effortless exercise in division (*short* division, assuming they still teach the 12-table in school), then one finds with remarkable ease:

$$\log 3 \approx 0.477139 \quad (\text{exact: } 0.477121\ldots), \tag{35}$$
$$\log 7 \approx 0.845136 \quad (\text{exact: } 0.845098\ldots), \tag{36}$$
$$\log 11 \approx 1.041421 \quad (\text{exact: } 1.041392\ldots). \tag{37}$$

Using (31) to improve the second level of approximation [Eqs. (21)–(24)] requires a little more effort, but it can still be done by hand in a page or two. The best students will be eager to try, and will be rewarded by

$$\log 2 \approx 0.30103015\ldots \quad (\text{exact: } 0.30102999\ldots), \tag{38}$$
$$\log 3 \approx 0.47712115\ldots \quad (\text{exact: } 0.47712125\ldots), \tag{39}$$
$$\log 7 \approx 0.84509790\ldots \quad (\text{exact: } 0.84509804\ldots), \tag{40}$$
$$\log 11 \approx 1.04139307\ldots \quad (\text{exact: } 1.04139268\ldots). \tag{41}$$

As a special treat for fanatics, the improvement can be applied to Eqs. (25) and (26). In fact less effort is required to do this by hand than to find the results (38)–(41). The only real burden is carrying to many places the long division by 1857, but for the pure of heart, the ineffable joy of seeing one after another of that amazing stretch of 9s come marching out of the deep interior of log 2, is more than enough reward:

$$\log 2 \approx 0.3010299987\ldots \quad (\text{exact: } 0.3010299956\ldots), \tag{42}$$
$$\log 3 \approx 0.4771212460\ldots \quad (\text{exact: } 0.4771212547\ldots). \tag{43}$$

After this it is almost too gross to observe that those who have fought through to these triumphant conclusions will be eager to be introduced to e itself, as $10^{0.43429\ldots}$; or that still finer results can be retained by introducing the quadratic term into Eq. (31); or that higher primes, at the very worst,[3] can be got at successively through $p^2 \approx (p-1)(p+1)$, since all prime factors of the right side must be less than p. Better, surely, to rest with the extraction of that glorious ...999.... And then get back to physics.

Acknowledgments

This paper was written while recuperating from surgery performed at Tompkins Community Hospital. I would like to thank an anesthesiologist who seemed to say that his name was Dr. Eeyore, for inducing the appropriate postoperative state of mind. I am also grateful to those many people whose names I do not know, who remembered to close the door when they left the room, and to N. W. Ashcroft, for a vase of lillies-of-the-valley.

Appendix

By taking the very small number X in

$$\log(1+X) \approx AX, \qquad |X| \ll 1, \tag{44}$$

to be the inverse of a very large integer n, one finds that the constant A should be given by

$$A \approx \log(1+1/n)^n, \qquad n \gg 1. \tag{45}$$

With the aid of the binomial theorem and a little rearranging, one has

$$\left(1+\frac{1}{n}\right)^n = 1+1+\frac{1}{2!}\left(1-\frac{1}{n}\right)+\frac{1}{3!}\left(1-\frac{1}{n}\right)\left(1-\frac{2}{n}\right) + \frac{1}{4!}\left(1-\frac{1}{n}\right)\left(1-\frac{2}{n}\right)\left(1-\frac{3}{n}\right)+\cdots. \tag{46}$$

From a few concrete examples one can persuade the amateur that when n is a *very* large number, then the first dozen or so terms in Eq. (42) will hardly depend on n at all. By the time one gets to terms that *do* depend appreciably on n they are so small (because of the inverse factorials) as to make negligible contributions to the sum. This not only provides a plausible basis for the original assertion that an approximation of the form Eq. (44) exists, but it also gives the numerical value of the constant:

$$A = \log(2+1/2!+1/3!+1/4!+\cdots) \tag{47}$$

as any professional knew all along.

The short computation of A starts from Eq. (47). If we need four place accuracy, it surely suffices to retain only terms through 1/7! in the sum. So truncated, the sum is most efficiently evaluated by giving each term the common denominator $7! = 5040$, which gives the result in the form 13700/5040. This fraction is more suggestively written as 2740/1008, not because it thereby bears a closer resemblance to the more familiar 2.71828..., but because, as the seasoned hunter of coincidences cannot fail to notice, it is thereby revealed on the 1% level of accuracy to be nothing but $3^3/10$. With the crudest estimate (12) for log 3, this gives

$$A \approx 3\log 3 - 1 \approx 17/40 = 0.425. \tag{48}$$

We can improve on this as follows:

More accurately,

$$A = \log\frac{2740}{1008} \approx 3\log 3 - 1 + \log\left(1 + \frac{4}{270}\right) - \log\left(1 + \frac{8}{1000}\right). \tag{49}$$

But by the nature of A Eq. (44) we can express the last two logarithms in terms of A again, to find

$$A\left(1 + \frac{8}{1000} - \frac{4}{270}\right) \approx 3\log 3 - 1. \tag{50}$$

To finish, we need merely repeat whichever of our calculations of log 3 strikes us as sufficiently accurate. If we take one of the estimates that exploited the correction formula (31), however, we must let the (momentarily unknown) symbol A replace the value 0.4343. Thus the procedure that gave Eqs. (34) and (35) now gives

$$10\log 2 \approx 3 + (24/1000)\,A, \tag{51}$$

$$4\log 3 \approx 3\log 2 + 1 + (1/80)\,A, \tag{52}$$

and therefore

$$\log 3 \approx \frac{19}{40} + \left(\frac{1}{320} + \frac{9}{5000}\right)A. \tag{53}$$

Using this to eliminate log 3 from Eq. (50) we find that

$$A\left(1 + \frac{13}{5000} - \frac{3}{320} - \frac{2}{135}\right) \approx \frac{17}{40}, \tag{54}$$

which a little scribbling shows to give

$$A \approx 0.43437\ldots \quad \text{(exact: } 0.43429\ldots\text{)}. \tag{55}$$

(Those whom it offends slightly to overshoot the second 43 can repeat the procedure with one of the better methods for getting at log 3, but for all the uses we have put it to, 0.4344 works just about as well.)

Notes and references

1 It is rather like the difference between Physics! and physics. If you know what I mean, please skip to p. 283.

2 Donald F. Holcomb and Philip Morrison, *My Father's Watch* (Prentice-Hall, Englewood Cliffs, NJ, 1974), pp. 107–112. Richard P. Feynman, Robert B. Leighton, and Matthew Sands, *The Feynman Lectures on Physics* (Addison-Wesley, New York, 1963), Vol. I, Sec. 22–4.

3 In any given case, one can almost always do better. One can, for example, get the logarithm of 13 through $7 \times 11 \times 13 = 1001$, or, if one is really serious, through $2^4 \times 3^4 \times 7^3 \times 11^3 \times 13^2 = 99991683792$. Afficionados should also note that if one plays the game on the primes through 13 one essentially gets 37 free of charge, since $3^3 \times 7 \times 11 \times 13 \times 37 = 999999$.

Postscript

"Logarithms!" produced the biggest correspondence of any paper I have written. Everybody wanted to tell me their favorite numerical discoveries. The nicest one I received was this: if you simply square 0.7, square the result, and keep squaring until you've done it nine times, you discover that

$$0.7^{2^9} = 4.9000046 \times 10^{-80} \tag{56}$$

so that

$$7^{512}/10^{512} \approx 7^2 \times 10^{-81} \tag{57}$$

or

$$7^{510} \approx 10^{431}. \tag{58}$$

The rational approximation this gives to log 7 all by itself without any further refinement is breathtaking.

23

Stirling's formula!

Stirling's approximation to the factorial function,

$$n! \sim (2\pi n)^{1/2}(n/\mathrm{e})^n, \tag{1}$$

plays a central role in any number of investigations of statistical physics, and is invaluable in the kinds of simple probabilistic studies that can convey to students in a general education course the nature of entropy and irreversibility. Unfortunately, the usual derivations of (1) are inaccessible to such students and even to many beginning physics majors. One can, of course, simply verify its remarkably accurate performance,[1] but the better students are bound to find this frustrating: Why is it that Stirling's formula works as well as it does?

I provide here an elementary answer to this question that can be adapted to give convincing explanations at a range of levels of mathematical innocence. For the crudest argument it is only necessary to know the elementary definition of the number e that arises in the theory of compound interest. Students who also know that the natural logarithm has the expansion

$$\ln(1+x) = x - x^2/2 + x^3/3 - x^4/4 + \cdots \tag{2}$$

can be given a really intimate glimpse into the workings of Stirling's formula, while those who are willing to approximate a few simple sums by integrals can acquire a level of understanding possessed, I suspect, by few professionals.

Stirling's formula begins to yield up its secrets with the observation[2] that $n!$ can evidently be written in the form

$$n! = (\tfrac{1}{2})(\tfrac{2}{3})^2(\tfrac{3}{4})^3(\tfrac{4}{5})^4 \cdots [(n-1)/n]^{n-1}n^n. \tag{3}$$

This can be written in the equivalent form,

$$n! = n^n \Big/ \prod_{j=1}^{n-1} \left(1 + \frac{1}{j}\right)^j, \tag{4}$$

which immediately calls to mind the definition of e as the limiting value for large j of $(1 + 1/j)^j$:

$$\mathrm{e} \sim (1 + 1/j)^j. \tag{5}$$

The denominator of (4) is thus a product of successively better and better approximants to e. If we were to estimate each term in the product by its limiting value e, we would extract from (4) the estimate

$$n! \sim n^n/\mathrm{e}^{n-1}. \tag{6}$$

For many purposes (6) is as good as one needs.[3] One can, however, do very much better if one realizes that although $(1 + 1/j)^j$ is indeed a good approximation to e for large j, the refinement

$$\mathrm{e} \sim (1 + 1/j)^{j+1/2} \tag{7}$$

does spectacularly better. The superiority of (7) to (5) is evident from Table I, and is easy to understand with the aid of the expansion (2). We have, on the one hand, the large j expansion

$$\begin{aligned} &(1 + 1/j)^j/\mathrm{e} \\ &\quad = \exp[j \ln(1 + 1/j) - 1] \\ &\quad = \exp[j(1/j - 1/2j^2 + 1/3j^3 - \cdots) - 1] \\ &\quad \sim \mathrm{e}^{-1/2j}, \end{aligned} \tag{8}$$

Table I. *The compound interest approximants to* e *(first column) and a much more rapidly convergent sequence (second).*

n	$(1 + 1/n)^n$	$(1 + 1/n)^{n+1/2}$
1	2	2.828...
10	2.59...	2.7203...
100	2.704...	2.71830...
1000	2.7169...	2.718282...
10000	2.71814...	2.71828183...
100000	2.718268...	2.718281828...

while, on the other hand, the ratio of the estimate (7) to e is given by

$$\begin{aligned}(1 + 1/j)^{j+1/2}/\mathrm{e} \\ &= \exp[(j+\tfrac{1}{2})\ln(1+1/j) - 1] \\ &= \exp[(j+\tfrac{1}{2})(1/j - 1/2j^2 + 1/3j^3 - \cdots) - 1] \\ &\sim \mathrm{e}^{1/12j^2}.\end{aligned} \qquad (9)$$

By stepping up the exponent by $\frac{1}{2}$, we have brought about the vanishing of the correction term in $1/j$, thereby reducing the error to one of order $1/j^2$ (with the agreeably small coefficient of $\frac{1}{12}$ appearing as a gratuitous bonus).

We can exploit the superiority of the estimate (7) of e by modifying (3) to the equally evident identity.

$$n! = (\tfrac{1}{2})^{1.5}(\tfrac{2}{3})^{2.5}(\tfrac{3}{4})^{3.5} \times (\tfrac{4}{5})^{4.5} \cdots [(n-1)/n]^{n-1/2} n^{n+1/2},$$

or, equivalently,

$$n! = n^{n+1/2} \Big/ \prod_{j=1}^{n-1} \left(1 + \frac{1}{j}\right)^{j+1/2}. \qquad (10)$$

If we now approximate each term in the denominator by e we arrive at the considerably improved estimate

$$n! \sim n^{n+1/2}/\mathrm{e}^{n-1}. \qquad (11)$$

This differs from Stirling's formula only in the replacement of $(2\pi)^{1/2} = 2.50662\ldots$ by $\mathrm{e} = 2.71828\ldots$, thereby overestimating the correct asymptotic form, but only by about $8\frac{1}{2}\%$.

The final step is suggested by writing the exact relation (10) as the approximate one (11) divided by the required correction:

$$n! = \frac{n^{n+1/2}}{\mathrm{e}^{n-1}} \Big/ \prod_{j=1}^{n-1} \left(\frac{(1+1/j)^{j+1/2}}{\mathrm{e}}\right). \qquad (12)$$

Note that when n is large the product is exceedingly insensitive to the actual value of its upper limit, since increasing n by one augments the product by a factor $(1 + 1/n)^{n+1/2}/\mathrm{e}$ which, for example, differs from unity by only a part in a billion when n is 10 000. Thus for large n we will do better if we estimate the product in the denominator not by unity, as in (10), but by its limiting value for infinite[4] n. To correct the error brought about by this extension of the product, we can introduce in the numerator the product of all the terms by which we have augmented the denominator, thereby maintaining the identity (12) but in the slightly modified form:

$$n! = \frac{n^{n+1/2}}{e^{n-1}} \frac{\prod_{j=n}^{\infty} [(1 + 1/j)^{j+1/2}/e]}{\prod_{j=1}^{\infty} [(1 + 1/j)^{j+1/2}/e]}. \tag{13}$$

The product in the denominator of (13) is some definite number C, independent of n; the product in the numerator depends on n, but approaches unity as n becomes large. If (13) is to agree with Stirling's formula (1) for large n, it must then be that

$$C = \prod_{j=1}^{\infty} \left(\frac{(1 + 1/j)^{j+1/2}}{e} \right) = \frac{e}{(2\pi)^{1/2}} = 1.084437551\ldots \tag{14}$$

If we insert this value into (13), with the understanding that π is some constant temporarily defined as $e^2/2C^2$, whose value we must eventually demonstrate to be 3.1415926535..., we arrive at the *exact identity*[5]:

$$n! = (2\pi n)^{1/2} \left(\frac{n}{e} \right)^n \prod_{j=n}^{\infty} \left(\frac{(1 + 1/j)^{j+1/2}}{e} \right). \tag{15}$$

This identity is the centerpiece of this note. It cannot be new, but knowledge of it seems to have been lost by current generations of physicists. I have failed to turn up any reference to it after a few hours hunting in the Mathematics Library, though I did find some people getting rather close in 1877.[6]

The celebrated success of Stirling's formula, even for small values of n, is now easy to account for. When $n = 1$ Stirling's formula falls short of the exact (15) by the product from 1 to infinity, which is just the constant $C = 1.0844\ldots$, whose closeness to unity is a measure of the excellence of the approximations (7) to e. When n is 2, the correction factor is the product from 2 to infinity – i.e., the worst of the estimates, $(1 + 1)^{1.5} = 2.828\ldots$ no longer appears, and the correction drops down to $Ce/2\sqrt{2} = 1.0422\ldots.$ With each additional step n takes through the integers that factor in the product farthest from unity drops out, yielding smaller and smaller successive percent corrections.

Two points remain to be made:

(a) The fact that the number π appearing in (15) has the value 3.1415926535... can be established in a suitably elementary manner.

(b) The exact form (15) can be used to give with minimum effort the leading terms in a series of improved approximations to $n!$

Point (b) is of no importance in any physical application I am aware of, but it is fun to show students how much better one can do with only a little more effort. Furthermore the results of such an extension can be put to use in establishing point (a). For this reason I attend first to point (b).

For the simplest estimate of the correction (15) gives to Stirling's formula, we need only use (9) to estimate the leading deviation of $(1 + 1/j)^{j+1/2}$ from e. This immediately gives as the correction factor

$$K = \prod_{j=n}^{\infty} \left(\frac{(1 + 1/j)^{j+1/2}}{\mathrm{e}} \right) = \exp\left(\sum_{j=n}^{\infty} \frac{1}{12j^2} \right). \tag{16}$$

Estimating the sum by the corresponding integral, we have at once:

$$n! \sim (2\pi n)^{1/2}(n/\mathrm{e})^n \mathrm{e}^{1/12n}, \tag{17}$$

whose spectacular improvement on Stirling's formula students should be urged to investigate for themselves.

Equation (17) does better than one might think it has any right to do. The reason is that (17) gets correct not only the leading correction factor $1 + 1/12n$, but also the next term in the series, $1/288n^2$. The full correction factor is, in fact, the exponential of a series in odd inverse powers of n. To see this and extract the next few terms as well, we write the full correction factor in the form

$$K = \prod_{j=n}^{\infty} \left(\frac{(1 + 1/j)^{j+1/2}}{\mathrm{e}} \right) = \exp\left\{ \sum_{j=n}^{\infty} \left[\left(j + \frac{1}{2} \right) \ln\left(1 + \frac{1}{j} \right) - 1 \right] \right\}. \tag{18}$$

The hidden structure of (18) is exposed by shifting from an integral variable of summation to one that assumes values halfway between the integers. Introducing the notational convention:

$$\sum_{n+1/2}^{\infty} f(m) = f(n+\tfrac{1}{2}) + f(n+\tfrac{3}{2}) + f(n+\tfrac{5}{2}) + \cdots, \tag{19}$$

and letting $j+\frac{1}{2}$ be a new variable m that assumes half-integral values, we rewrite (18):

$$K = \exp\left\{\sum_{n+1/2}^{\infty}\left[m \ln\left(\frac{1+1/2m}{1-1/2m}\right) - 1\right]\right\}. \tag{20}$$

Expanding separately the logarithms of $1 \pm 1/2m$ then gives

$$K = \exp\left[\sum_{n+1/2}^{\infty}\left(\frac{1}{3(2m)^2} + \frac{1}{5(2m)^4} + \frac{1}{7(2m)^6} + \cdots\right)\right]. \tag{21}$$

By introducing the half-integral summation variable we have thus disposed of the odd terms in the expansion of the logarithm. But this is not the only service it performs for us. It is also tailor made for estimating the error we make when we go on to approximate the sums in (21) by integrals. To see this, start with the elementary identity

$$\int_n^{\infty} f(x)\,\mathrm{d}x = \sum_{n+1/2}^{\infty} \int_{-1/2}^{1/2} f(m+x)\,\mathrm{d}x, \tag{22}$$

and expand each $f(m+x)$ about $x=0$:

$$f(m+x) = f(m) + xf'(m) + (x^2/2)f''(m) + (x^3/6)f'''(m) + \cdots. \tag{23}$$

Note that only the even terms survive the integration over the symmetric interval $(-\frac{1}{2}, \frac{1}{2})$. The leading term gives the sum we wish to evaluate; the remaining terms give corrections to its approximation by the integral:

$$\sum_{n+1/2}^{\infty} f(m) = \int_n^{\infty} f(x)\,\mathrm{d}x - \sum_{n+1/2}^{\infty} \frac{1}{3!} f''(m)/2^2$$
$$- \sum_{n+1/2}^{\infty} \frac{1}{5!} f^{iv}(m)/2^4 - \sum_{n+1/2}^{\infty} \frac{1}{7!} f^{vi}(m)/2^6 - \cdots. \tag{24}$$

The expansion (24) can be used to evaluate the sums in (21) to give corrections to (17) beyond the leading term $1/12n$. Suppose, for example, we desire the correction up to and including terms of order $1/n^5$. Then the terms beyond $1/m^6$ in (21) can be ignored, and we can estimate the m^{-6}, m^{-4}, and m^{-2} terms successively as follows:

We only need the leading term in (24) to establish to the required degree of accuracy that

$$\sum_{n+1/2}^{\infty} \frac{1}{m^6} \sim \int_n^{\infty} \frac{dx}{x^6} = \frac{1}{5n^5}. \tag{25}$$

The term in m^{-4} requires the first two terms in (24):

$$\sum_{n+1/2}^{\infty} \frac{1}{m^4} \sim \int_m^{\infty} \frac{dx}{x^4} - \frac{5}{6} \sum_{n+1/2}^{\infty} \frac{1}{m^6} = 1/3n^3 - 1/6n^5, \tag{26}$$

where we have used the result (25) to evaluate the sum of m^{-6}. Finally, the term in m^{-2} requires the first three terms in (24):

$$\sum_{n+1/2}^{\infty} \frac{1}{m^2} \sim \int_n^{\infty} \frac{dx}{x^2} - \frac{1}{4} \sum_{n+1/2}^{\infty} \frac{1}{m^4} - \frac{1}{16} \sum_{n+1/2}^{\infty} \frac{1}{m^6}$$
$$= 1/n - 1/12n^3 + 7/240n^5, \tag{27}$$

where we use the results (25) and (26) to evaluate the sums of m^{-4} and m^{-6}.

Using these three results to evaluate the first three terms in the expression (21) for the correction factor K to Stirling's formula leads immediately to

$$n! \simeq (2\pi n)^{1/2} \left(\frac{n}{e}\right)^n \exp\left(\frac{1}{12n} - \frac{1}{360n^3} + \frac{1}{1260n^5}\right), \tag{28}$$

where the error is now of order $1/n^7$ in the argument of the exponential. It is the rare student who is not deeply moved by the results of testing (28) on a programmable pocket calculator for successive values of n, starting with unity. I know of no more vivid demonstration of the power of a very little bit of elementary analysis.

But what about the pi? One could, of course, determine the constant in (28) to a precision well beyond anything one would ever require by, for example, fitting it so that (28) gave $n!$ exactly for $n = 10$ (which gives π to ten place accuracy). I find the following approach, however, more adaptable to numerical computation, more esthetic, and more in line with the original statistical motivation for the whole investigation.

Consider the quantity

$$P = (2n)!/2^{2n}(n!)^2. \qquad (29)$$

This is, of course, the probability of getting an exact 50–50 split in $2n$ tosses of a balanced coin. By multiplying out everything in (29) we find that

$$P = \frac{1 \cdot 3 \cdot 5 \cdot 7 \cdots (2n-1)}{2 \cdot 4 \cdot 6 \cdot 8 \cdots (2n)}. \qquad (30)$$

On the other hand, using (28) to estimate the value of P for large n gives

$$P \sim \frac{1}{\sqrt{\pi n}} \exp\left(-\frac{1}{8n} + \frac{1}{192n^3} - \frac{1}{640n^5} \right). \qquad (31)$$

Students who know about Wallis's product for pi,

$$\frac{\pi}{2} = \frac{2}{1} \cdot \frac{2}{3} \cdot \frac{4}{3} \cdot \frac{4}{5} \cdot \frac{6}{5} \cdot \frac{6}{7} \cdot \frac{8}{7} \cdots \qquad (32)$$

can be immediately persuaded by a comparison of (31) (*without* the exponential correction) and (30) that our pi is *the* pi. Those who do not could be invited to evaluate the Wallis product numerically, but its convergence is exasperatingly slow. It is better to use the correction factor in (31) to estimate the unevaluated remainder in Wallis's product:

$$\pi = 4 \cdot \frac{2}{3} \cdot \frac{4}{3} \cdot \frac{4}{5} \cdot \frac{6}{5} \cdot \frac{6}{7} \cdot \frac{8}{7} \cdots \frac{(2n-2)(2n)}{(2n-1)(2n-1)}$$
$$\times \exp(-1/4n + 1/96n^3 - 1/320n^5). \qquad (33)$$

This produces values of π well beyond anything most physicists would care to remember. (For $n = 10$, 3.1415926528...; for $n = 50$, 3.141592653589784....)

There are, of course, students who will want to know *why* Wallis' product should have anything to do with the circumference of a circle – indeed, they are the ones whose persistent questioning inspired this essay in the first place, and they deserve a better answer. For them the approach I prefer – the one I find (dare I say it?) the most *physical* – is to let the answer emerge in the subsequent investigation of entropy.

Elementary insights into irreversibility can be had from studying the statistics of coin tossing. For this one needs to know not only the probability (29) of an even break in $2n$ tosses, but, more generally, the probability of the number of heads or tails deviating from n by an amount x. When n is large one can use Stirling's formula (without the correction factor, but with our undetermined pi) to evaluate the appropriate binomial coefficients, and convince the class that the resulting expression is Gaussian when x is of order $n^{1/2}$, and utterly negligible when x is any greater. Determining the value of *our* pi then reduces to the problem of normalizing a Gaussian distribution, which is done in the usual way, squaring the normalization integral and evaluating it in polar coordinates. *The* pi comes from the angular part, the deed is accomplished, and the course can march on into the study of irreversibility, securely based on the rock solid foundation provided by Stirling's self-evident approximation to the elementary identity (15).

Notes and references

1 See, for example, S. A. Feller and E. Kaspar, *Am. J. Phys.* **50**, 682 (1982); or Y. Weissman, *Am. J. Phys.* **51**, 9 (1983).

2 Although this article addresses teachers, not students, I have sketched the argument along lines one might present to a relatively innocent class, relegating more learned asides to footnotes. Readers in a hurry should skip directly to Eq. (15) and its accompanying note.

3 The approximation to $n!$ is poor, since cumulative errors lead to a divergent correction factor, as remarked upon in footnote 4, below. The approximation to $\ln(n!)/n$, however, gets quite good as n increases since it averages over a set of increasingly good approximants to $\ln(e)$.

4 That the infinite product converges is an immediate consequence of the expansion (9), since the sum of $1/j^2$ converges. Note that this convergence would fail had we tried to base an analogous argument on the inferior estimate (8), owing to the divergence of the sum of $1/j$. The degree to which this entire point should be soft-pedaled in a general education course is best left to the discretion of the instructor. It would never occur to many such students that the infinite products in (13) *could* diverge, but the morality of relying on such innocence must be weighed in the conscience of each pedagogue.

5 For the sophisticate who already knows Stirling's approximation it should be immediately evident that (15) is exact: (a) the product converges as a direct consequence of (9); (b) the value of (15) for $n+1$ is directly and easily shown to

be $n+1$ times its value for n so the entire expression is indeed a constant times $n!$, and (c) the constant is determined by thc requirement that (15) agree with (1) in the large n limit.

6 See J. W. Glaisher, *Q. J. Math.* **15**, 57 (1877), and note the comment by Cayley at the end.

24

Pi in the sky

H. Aspden's[1] amusing "unbelievable" formula for $g/2$ reminds me of Ramanujan's remarkable number $(2143/22)^{1/4} = 3.14159265258\ldots$ (to be compared with $\pi = 3.14159265358\ldots$).

The agreement with π is to 300 parts in a trillion, a factor of 10 less spectacular than Aspden's agreement with quantum electrodynamics (QED), but, like his, still well within the error of even the most careful measurements of the circumference and diameter of the very finest physical circles. Should one hope that a future theory will confirm that π is indeed equal to $(2143/22)^{1/4}$ thereby freeing us from the need to compute higher and higher order terms in any of the many tedious series mathematicians have had to resort to in these unenlightened times?

How "unbelievable," in fact, is Ramanujan's formula? I would say it's surprising, but not astonishing – the odds against it are certainly no more than 1000 to 1. In support of this claim, ask yourself how Ramanujan could possibly have discovered such a formula. One way is by noticing that the continued fraction expansion for π^4 has the form

$$\begin{aligned}\pi^4 = 97 + 1/(2 + 1/(2 + 1/(3\\ + 1/(1 + 1/(16539\\ + 1/(1 + 1/(10\\ + \ldots))))))).\end{aligned}$$

[1] H. Aspden, *Am. J. Phys.* **54**, 1064 (1986). Aspden says there is a theoretical model behind his formula, but he invites the reader to consider the result in and of itself, "without further elaboration" and it is in that spirit that I offer this reaction.

The miracle lies in the appearance of that enormous denominator 16539 so early in the expansion. If you replace 16539 by ∞, you get Ramanujan.

I'm no expert on the statistics of continued fraction denominators, but I would put the odds against such a big one so soon at about 10000 to 1. (Big ones aren't *that* rare: a big one – in the hundreds – is responsible for the excellent approximation to π, 355/113.)

Another way to rediscover Ramanujan's relation is to note that $\pi^4 = 97.409091034002\ldots$, which is a pretty good imitation of $97.40909090909\ldots = 2143/22$. Given the first 09 in π^4 (which could have been anything else) the chance of its immediately repeating and then almost (10 instead of 09) repeating again are just about 3000 to 1 – the same kind of number.

But then what's special about the 4th root? Surely a formula involving any reasonably low root – say 12th or less – would have been just as noteworthy, so that reduces the odds by another factor of 10. It's hard to know how amazed one should be at the fact that only the digits 1, 2, 3 and 4 appear in the formula and in a rather pleasing arrangement at that. Nobody can deny that this is part of its magic, but then again almost any group of seven digits has something striking about it or I, for one, could never remember anybody's telephone number.

Ramanujan's approximation to π offers a rather more transparent example of an unbelievable formula against which to assess the validity of one's amazement at Aspden's. Speaking for myself, I'm a little surprised. But not at all astonished.

25

Variational principles in dynamics and quantum theory

By Wolfgang Yourgrau, Stanley Mandelstam, Saunders, Philadelphia, 1968.

This is the third edition of a little volume, first published in 1952, that traces the uses of the variational principle in physical science all the way from Hero of Alexandria to Schwinger of Massachusetts. The occasion for the new edition is the addition of an essay by Laurence Mittag, Michael Stephen, and Wolfgang Yourgrau on the hydrodynamics of normal fluids and superfluids, emphasizing variational formulations.

The book, however, is worth commenting on as a whole, for it has languished in an undeserved obscurity. It offers concise and elegant formulations of the great variational principles of optics, mechanics, electrodynamics, quantum theory (old and new), and now, hydrodynamics. Although each topic is skillfully presented in analytic terms congenial to the modern reader, the historical background is almost always preserved. This is a luxury almost universally abandoned by authors of scientific books, and though the reasons for this are obvious, the loss is quite possibly greater than many of us think. It is not just that the subject is thereby dehumanized; the accompanying loss of perspective on our own efforts, both as physicists and as people, is equally to be regretted. A trivial example is the discussion of Hamilton–Jacobi theory (the concise presentation of which, by the way, is the best I have seen this side of Landau and Lifshitz) which treats us to the spectacle of Jacobi, unable to resist the needle, in presenting his modest refinement of Hamilton's remarkable edifice (page 58): "I therefore do not know why Hamilton ... requires the introduction of a function S of $6n+1$ variables ... while, as we have seen, it is completely sufficient to ...." Does not that blend of pride and false innocence recall innumerable referee's reports, replies to referee's reports, and now *Physical Review* comments and counter-comments that we have all read and (alas) written? Compare the dignity of Hamilton on Lagrange (page 44): "Lagrange has perhaps done more than any other analyst, to give extent and harmony to such deductive researches, by showing that the most varied consequences respecting the motions of systems of bodies may be derived from one radical formula; the beauty of the method so suiting the dignity of the results as to make his great work a kind of scientific poem."

Would that we are all in dread of being remembered as much for our prose as our theories and experiments; it might well benefit all three.

Not the least of the pleasures offered by Mandelstam and Yourgrau is the abundance of nuggets like the two cited above. A few more examples:

Whittaker on Hilbert's variational formulation of Einstein's law of gravitation: "Gravitation simply represents a continual effort of the universe to straighten itself out."

Poisson on the principle of least action (1837): "... only a useless rule."

Planck on the principle of least action (1915): "... that [general law] which ... may claim to come nearest to that ideal final aim of theoretical research."

My complaints are few. Several philosophical axes are ground at regular intervals throughout the text, which process I personally found of little interest. Furthermore (and I offer this comment with the greatest reluctance in this day when scientific prose has completed its evolution into a medium with the vibrancy and resilience of wet cotton) I was continually disconcerted by a prose style I can only describe as baroque (example: "Almost overbearingly did he postulate...") and a staggering lack of humour. Two examples of the latter:

Could not Planck have been joking when he viewed "as a

variational principle Leibniz's maxim that our world is the best of all possible worlds!"

Could Voltaire (of all people), speaking of Maupertuis, not have been joking when he "dubbed him 'Sir Isaac Maupertuis,' so elevating him to the lofty rank of Newton," especially in view of the fact that he "later inveighed against his former idol with uncontrolled invective."

If not, your reviewer stands convicted of a most unseemly frivolity and irreverence. Be that as it may. The book as a whole should be a delight to the educated and a most valuable supplementary text for the student.

26

Special functions: a group theoretic approach

By James D. Talman,
Benjamin, New York 1968.

This delightful volume does for most of one's favorite special functions what arithmetic does for the exponential: it displays them not as solutions to differential equations but as matrix elements of representations of elementary Lie groups.

If you feel deeply that there is nothing more to say after pointing out that e^x is the solution of $f' = f$ (with $f(0) = 1$) and do not care that it is also 2.71828 ... multiplied by itself x times, then you also might feel inclined to ignore this book. This would be a pity, because almost half is devoted to the kind of general exposition of Lie groups, Lie algebras and their representations that few mathematicians are capable of treating us to. It is an exposition that prefers words to symbols, is intuitive and does not require us to scale a mountain when we only want to peek over the garden hedge. Knowing only a little linear algebra, one can read this much with joy and enlightenment without compromising one's devotion to differential equations.

If, however, you have always suspected that there was more to Bessel functions, Gegenbauer polynomials, associated Laguerre polynomials, and Hermite polynomials than Morse and Feshbach or Watson were letting on, then you can indulge in the feast as well as the cocktails. And there, in the algebraic paté, you will come upon those higher transcendental truffles you long ago snuffled in the garden of analysis. Only a hard man could fail to be moved by this transformation, and you can see what it has done to a soft-hearted esthete like your reviewer.